SOLAR ENERGY
EXPERIMENTS

For High School and College Students

SOLAR ENERGY EXPERIMENTS

For High School and College Students

Thomas W. Norton *Author*

Research Associate, Atmospheric Sciences Research Center
State University of New York at Albany; Physics Instructor,
Linton High School
Schenectady, New York

Donald C. Hunter, P.E. *Consultant*

Chief, Manual Air Monitoring Section
New York State Department of Environmental Conservation
Albany, New York

Roger J. Cheng *Project Coordinator*

Research Associate, Atmospheric Sciences Research Center
State University of New York at Albany

Rodale Press   Emmaus, PA

2 4 6 8 10 9 7 5 3

Library of Congress Cataloging in Publication Data

Norton, Thomas W.
   Solar energy experiments for high school and college students.

   Bibliography: p.
   Includes index.
   SUMMARY: Presents eighteen self-explanatory solar energy experiments and classroom activities suitable for individual student analysis.
   1. Solar energy—Experiments—Juvenile literature.
[1. Solar energy—Experiments] I. Title.
TJ810.N67   621.47   77-4918
ISBN 0-87857-179-5

Printed in the United States of America on recycled paper

CONTENTS

FIGURES AND TABLES

EXPERIMENTS

CLASSROOM ACTIVITIES

Contents

ABSTRACT

This is a multipurpose manual comprising eighteen experiments and eight classroom-type activities. The eighteen experiments are of varying difficulty and cover the important aspects of solar energy utilization. Each experiment is self-contained, with its own introduction and background information, making this book a suitable solar energy curriculum outline.

Energy measurements are emphasized, and techniques for collector efficiency determinations are considered. Among the topics discussed are altitude and azimuth of the sun, radiation characteristics, energy collection with converging lenses, air and water solar collectors, and energy storage in gravel beds and in salt hydrates. Both theoretical and practical engineering considerations are illustrated by the experiments. Many experiments are directly applicable to existing physics, general science, and environmental science curricula, while others are of sufficient difficulty and duration to challenge college students and the most advanced secondary students.

The eight classroom-type activities present worldwide energy data and solar energy data for individual student analysis. As a result, this manual can serve as a useful classroom resource as well as a general reference source.

PREFACE

The New York State Legislature appropriated $200,000 to the Atmospheric Sciences Research Center for research and education in solar energy. This special appropriation was for the budget period 1 April 1974 to 31 March 1975. The legislators expressed their confidence that government and science can develop a meaningful partnership to help resolve the energy crisis in New York State. Solutions to the energy crisis require action on well-coordinated, long-range programs, particularly in the fields of alternate energy sources and energy conservation.

The Atmospheric Sciences Research Center has initiated broad research and educational programs in these areas, such as:

1. Data collection and evaluation for the preparation of a detailed atlas on the available solar energy in New York State.

2. Construction of a small "Minimum Energy Building" utilizing advanced insulation material in conjunction with solar energy systems for heating and cooling. A complete systems analysis will be performed.

3. Participation in the construction of a solar heating and cooling demonstration project. The Alumni House (a 5,000-square-foot conference center for the State University of New York at Albany) will be instrumented and studied to assess the impact of solar energy in the upper New York State area. A complete systems analysis will be performed.

4. Research into heat storage materials. The concept of decreasing the heat storage volume or increasing the efficiency of heat storage per unit volume has considerable economic impact on initial and operating costs for the utilization of solar energy.

5. Construction and evaluation of solar-heated greenhouses. Assessment of the economic and the technological feasibility of these structures in New York State, particularly in the northern regions. This project is directed by the State University of New York at Plattsburgh and the Miner Institute.

6. Development of a predictor model allowing the assessment of the effect of air pollution on the amount of solar energy available at the ground. This information will be incorporated into the computerized performance modeling of future solar heating and cooling construction.

7. Development of a solar technology curriculum for two-year colleges. This will include academic and technological programs in cooperation with the Adirondack Community College.

8. Publication of a laboratory manual entitled *Solar Energy Experiments for Secondary School Students,* in cooperation with the New York State Department of Environmental Conservation and Linton High School, Schenectady, New York.

9. Standardization of calibration for solar radiation. Development of portable solar radiation detectors allowing intercomparison between all instruments presently operational in New York State. Fixed instrumentation located at the Schenectady County Airport will be accessible to the private and industrial sectors for comparison studies.

10. Development of a solar station at Whiteface Mountain summit which will transmit all data obtained to the base station, and from there data will be linked to the Department of Environmental Conservation's central computer station.

The Atmospheric Sciences Research Center and its professional staff are committed to the implementation of alternate energy sources in New York State. In conjunction with proper energy conservation programs, solar energy will aid us in continuing the economic well-being of the state with reduced reliance on out-of-state sources of nonrenewable products such as petroleum and coal. The efforts that lie ahead of us are monumental. The appropriation by the legislature of the state of New York, the first of its kind in the United States, marks the beginning of a development that will eventually remove the economic, social, and environmental constraints imposed upon each individual by the ever-increasing scarcity of fossil fuels.

Volker A. Mohnen
Project Director for the
Solar Energy
Appropriation and
Director
Atmospheric
Sciences
Research Center

INTRODUCTION

No one knows for sure just how long our fossil fuel resources of coal, oil, and natural gas will last. But we do know they are finite. And we know they are being consumed at an ever-increasing rate. We can't wait until they are gone before making provisions for use of alternate types of energy. That's why it is so important for us to learn all we can about energy sources which are not finite. The sun provides unlimited energy, and it is clean; that is, it doesn't produce environmental pollution. We can stretch fossil fuel resources by taking every possible opportunity to use clean energy from sources such as the sun, the wind, and moving water. Other sources of energy, unknown now, may be discovered and developed in the future.

We should adopt the attitude that we are stewards of the earth and of our environment during our sojourn here, and should consider that we have a responsibility far beyond our own all-too-short lives. Just as parents provide for the physical, mental, and emotional well-being of their children, so must we be concerned about our grandchildren, great-grandchildren, and future descendants by leaving a good environment with energy resources which can help make their lives worth living.

This manual is intended to act as an "idea seeding" source so that high school and college students and teachers can become more familiar with the solar energy aspects of the energy problem. We hope that the experiments will help the interested student relate his science classroom work to the practical uses of solar energy and encourage his independent experimental work. The activities are offered so that a teacher can get the whole class involved. This should help the nonexperimentally oriented students become aware of their own energy consumption, and perhaps encourage their energy conservation efforts. One important outcome should be a greater feeling for the order of magnitude of the energy problem—certainly on the personal level. By extrapolation, perhaps the overall problem can come into focus.

The experiments provide such a frame of reference. Some are fairly straightforward, but others will require the ingenuity and resources of the student; this is what makes experimenting fun. Some have not been field tested. As a result, I would be particularly pleased to receive comments regarding their use and also any data that are collected. Suggestions for improvement and ideas for additional experiments and activities will be most welcome. In turn, I would be interested in helping students who might have experimental difficulties.

Thomas W. Norton
Linton High School
The Plaza
Schenectady, New York
12308

OVERVIEW—MEASURING SOLAR ENERGY

The altitude and azimuth experiments measure the geometric variation of the sun's rays at any point on the earth's surface. The purpose of these experiments is to relate this variability to the actual energy received at your observing point on the earth's surface. By measuring the solar radiation, you will be able to perceive the transfer of energy from the sun to the earth.

One way to measure the energy being received by a body is to determine the effect the energy has on the body when it is absorbed. One such effect is that it raises the temperature of the receiving body. Thus, the thermometer becomes an important instrument in measuring solar energy. It does *not* measure heat energy directly, but does tell us if any energy absorbed causes an increase in the speed of the molecules of the absorbing body. If it does, the energy absorbed can be measured from the temperature change that results.

Should the absorbed energy cause the molecules of the body to separate from each other and get farther apart (in most cases) without an increase in temperature, a phase change is said to have occurred. This is what happens when a solid changes to a liquid, or a liquid changes to a gas. In each case energy is required to do this, and the amount of this energy (latent heat) depends on the amount of material (mass) the body has and the kind of material that is undergoing the phase change. Tables of latent heats are published for most materials, so that the measurement of the heat absorbed becomes one of determining the mass of material that has changed phase.

Another effect of incident radiation on certain bodies is the release of electrons from their surface. Some materials are very photosensitive in this way. The number of electrons that are freed from their atoms can be used as a measure of the amount of energy being received. This can be measured as an electrical current with an ammeter.

Let's try some experiments that demonstrate each of these techniques. Remember, however, that three things happen at a surface receiving radiation: reflection, transmission, and absorption of the radiation. These experiments will only measure the effects of the absorbed radiation, so care must be taken in interpreting your results. You cannot say that the quantity of energy you measure is the quantity of energy incident upon the body. But you can compare your measurements with published values of incoming solar radiation at your latitudes to obtain some idea of your measuring efficiency. In each case, the efficiency of the energy conversion becomes an important factor.

Rather than taking a definite stand on using only the metric system or the English system of units, we have used both in the way both are encountered in the field. The more theoretically oriented experiments favor the metric system, while those that involve construction and/or engineering viewpoints favor the English system. It is well for students to learn to feel comfortable with both when becoming involved with current solar energy practices.

Caution

A note of caution is appropriate for anyone attempting to work with solar radiation. The sun's intensity is great enough to seriously damage the retinas of your eyes when you look at the sun directly or at its image when concentrated by converging lenses. Fire hazards exist when working with concentrated rays, and solar collectors have been known to overheat, boil over, or otherwise spoil the day for the unwary experimenter. Good laboratory safety procedures must be practiced by all who choose to work in this field.

Determining the Altitude
and Azimuth of the Sun

Objectives

1. To measure the altitude of the sun.
2. To measure the azimuth of the sun.

Introduction

This is perhaps the most fundamental of all solar energy experiments since it defines the behavior of the sun from which we expect to obtain energy. The determination of the sun's position in the sky at a given location on the earth's surface is important to our understanding of the varying amounts of solar energy we receive, as well as to our most efficient use of that energy.

You have observed the sun low in the eastern sky in the morning, highest in the sky at noon, and low in the western sky in the evening. This apparent motion of the sun is due to the axis rotation of the earth. Perhaps you have also observed how high the sun is at noon in early summer, while at noon in early winter it is much lower in the sky. This apparent change in the north-south position is due to the tilt of the earth's axis and the earth's revolution around the sun once each year. Figure 1–1 shows the position of the sun and its path across the sky for various times of the year. The shadows

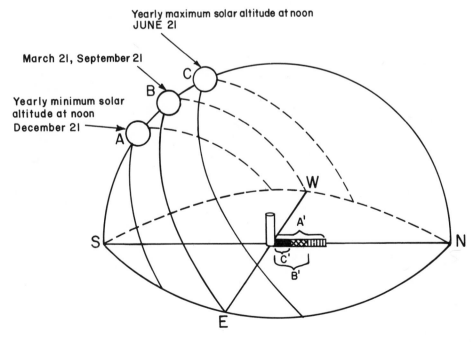

Yearly maximum solar altitude at noon
JUNE 21

March 21, September 21

Yearly minimum solar
altitude at noon
December 21

Figure 1—1
Various paths of the sun across the sky, and the corresponding shadow lengths at noon on the equinoxes and summer and winter solstices.

1

cast by a vertical post are also shown for corresponding sun positions. You can see that only twice a year does the sun rise due east of an observer and set due west. This occurs on the vernal equinox (March 21st) and on the autumnal equinox (September 21st). You can see from figure 1–2 that when the sun is highest in the sky on any day, the shadow cast at that time is shortest and the shadow line is aligned due north. You can check this by comparing the alignment with the North Star (Polaris) in an evening observation.

Figure 1—2
Various shadow positions and lengths during the summer solstice.

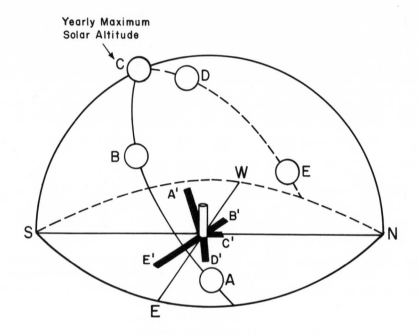

Yearly Maximum
Solar Altitude

You have also observed that temperatures are lower when the sun is low in the sky and are higher when the sun is high in the sky. This is because the heating effect of the sun's rays are the greatest when the sun's rays are direct—that is, they are striking a surface at an angle of 90°. When the rays strike a surface at small angles, the same rays cover a much larger area, so that each unit area of surface receives less energy.

When the rays are striking the surface at 90°, the angle of incidence is said to be zero, since, by convention, incident angles are measured with respect to a normal (perpendicular) to the surface at the point of incidence. When the angle of incidence is zero, the heating effect of the insolation (incoming solar radiation) is at a maximum, and when the angle of incidence increases to 90°, the absorbed insolation decreases to zero. Thus, the intensity of solar radiation on a surface varies with the cosine of the incident angle of the sun's rays. For this reason, the effectiveness of any solar energy device will depend on its alignment with the sun. For the geometric analysis showing these relationships, see page 7. In general, horizontal collectors will receive less energy than correctly tilted collectors, and fixed south-facing collectors will receive less energy than those that can follow the sun across the sky. However, any movable collector will require rather complex mounting that would add greatly to the initial expense and upkeep, so frequently the fixed collectors make more economic sense.

Another effect serves to decrease the absorbed insolation. When the sun is lower in the sky, the sun's rays must travel a greater distance through the atmosphere before reaching the earth. This tends to reduce the heating effect at the surface since a greater portion of the radiation will be absorbed by the atmosphere than when the sun is high.

So, you can now see why we want to measure how high the sun is in the sky and to keep track of its position. The effects of the apparent motion of the sun will appear in the results of other solar energy experiments. The sun's height in the sky, measured in degrees from the horizontal, is called the sun's altitude. The horizontal direction, measured in degrees from the true south line to a point on the surface directly beneath the sun, is called the solar azimuth. This definition differs from that used by surveyors, who consider azimuth as the horizontal angle in degrees clockwise from true north to the vertical projection of the line of sight. See figure 1–3.

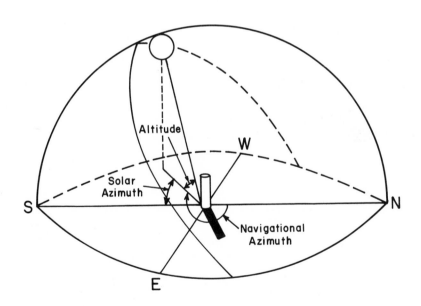

Figure 1—3
Solar azimuth, navigational azimuth, and altitude of the sun.

Three-inch finishing nail, 1-square-foot sheet of ½-inch plywood, 10-×-10 graph paper, straight edge, protractor, hammer, millimeter scale, carpenter's tri-square, level.

Materials

Procedure 1   Measuring the sun's altitude (vertical angle).

1. Scribe a straight center line across the plywood sheet and drive the finishing nail vertically into the plywood on this line near one end. Check the vertical position of the nail, using a square or right triangle.

2. At 12 o'clock noon, standard time, align the plywood on a *level* surface so that the nail is toward the south and so that the nail's shadow falls on the scribed line. The scribed line will then be in a nearly north-south direction. Fasten the plywood to its support and make sure all the measurements you make throughout this experiment are taken with this same orientation.

For a more precise determination of a true north-south line, make a four-minute correction to your noon (clock time) for each degree of longitude difference between your observation position and the center of your time zone. For example, if your longitude is 78° west, the sun would cross your meridian (north-south line) 12 minutes later than it did the 75th meridian, which is the center of the eastern time zone. If your longitude is 73° west, the correction is −8 minutes, and the sun would be on your meridian at 11:52 o'clock. In addition, corrections could be made for the fact that the sun is sometimes ahead and sometimes behind the clock depending on where the earth is in its revolution around the sun (due to changing speed). This latter correction could amount to 15 minutes between local solar noon and 12 noon standard time. The equation of time indicates the appropriate correction (fig. 1–4). Correction for daylight saving time should be made, too.

A simpler experimental method would be to note the alignment of the shadow when it is shortest. It would then fall on your meridian and the time would be local solar noon.

3. By measuring the height of the protruding nail and the length of the nail's shadow, using the millimeter scale, you will have two measurements that can tell you the sun's altitude at that time.

The altitude is the angle between the horizontal surface and the line of sight to the sun. A line connecting the top of the nail and the tip of the shadow will point directly to the sun and form a right triangle with the nail and shadow. The angle it makes with the horizontal board is the altitude of the sun. From simple trigonometry, you can see that the tangent of this angle is equal to the height of the nail (opposite side) divided by the shadow length (adjacent side). By making this computation and referring to trigonometry tables, you can obtain the angle. An equally valid way to do it would be to construct two perpendicular lines, measure off the length of the shadow on one line, the height of the nail on the other, connect the ends of the lines with a third line (the hypotenuse of the right triangle), and use a protractor to measure the altitude, directly from this

4

Minutes

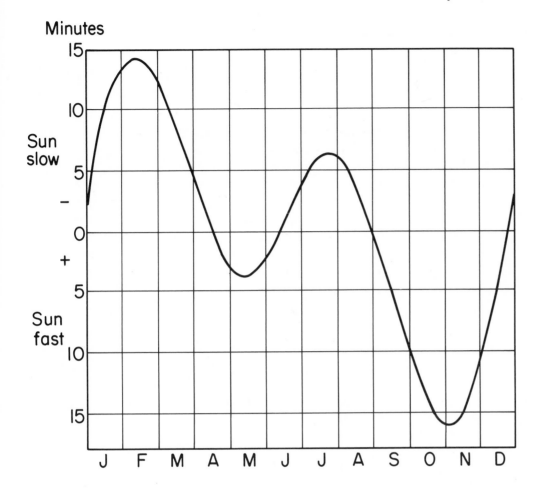

Figure 1—4
Equation of time throughout the year. (After E. A. Fath, *Elements of Astronomy*, Mc-Graw Hill, 1934.)

plot. The angle between the hypotenuse and the shadow length is the altitude of the sun in degrees.

4. By making measurements at 9 A.M., noon, and 3 P.M. (or at other hours if you wish) on or about the 21st day of each month, you will have a good yearly record of the sun's altitude on days coinciding with the solstices and equinoxes.

Measuring the azimuth (horizontal angle).

Procedure 2

1. Every time you measure the length of the shadow for your altitude measurements, measure the angle that the shadow makes with the scribed north-south line. This shadow angle is used to determine the solar azimuth. You can see from figure 1–3 that this angle is the vertical angle of the solar azimuth and hence is equal to it.

2. It would be instructive for you to actually construct these lines on plain or graph paper. Draw a line representing your meridian. Draw another line intersecting this at the shadow angle that you measured. The intersection of these lines represents the position on the earth from which

5

you made your observations. Note that the shadow itself is a horizontal projection of a part of the sight line onto the surface.

3. Extend this shadow line through the observing point toward the sun's position. This extension is the horizontal projection of the sight line to the sun, and the angle it makes with your meridian is the solar azimuth.

4. You will note that the shadow falls to the west of your meridian in the morning when the sun is in the eastern sky. There will be a corresponding sun-shadow position in the afternoon making the same angle with the north-south line. Thus, solar azimuths have two corresponding sun positions each day. For a 10° azimuth, there is a morning and afternoon position. Disregarding weather and atmospheric conditions, the insolation characteristics at your location in the afternoon will be a mirror image of those found in the morning. Record the altitude, date, and time of day for each observation.

5. On the equinoxes, you will observe that the sun rises due east and sets due west of your position. At these times the solar azimuths are 90°. Between the spring and autumn equinoxes, the early morning and late afternoon shadows fall south of an east-west line since the sun rises and sets north of the east-west line. The solar azimuth is measured relative to the north part of the north-south line at these times. Thus you can see the need to make careful records of which reference lines you use for your azimuth determinations.

The azimuth system used generally in the sciences, as well as for navigation and military purposes, numbers the degrees clockwise from 0° at the north point through a full circle of 360°. Here, no letters are needed and no numbers are repeated. Also, the addition and subtraction of azimuth angles are quite simple. While this system presents fewer difficulties in this way, the symmetry of azimuth angles about the north-south line in the morning and afternoon is not so obvious. Most solar energy references do not use this navigational system for this reason.

Results

1. Plot the values of altitudes you obtained versus the time of day for each month on a single graph. A family of curves will be obtained that will show the variations in the altitude of the sun that occur daily and yearly at your location.

2. A similar family of curves may be made for your solar azimuth data.

3. A third curve of shadow length versus altitude of the sun would give you a nice calibration curve for your instrument for determining the altitude directly from your shadow measurement.

Save these results so that you can compare them with the results of some other experiments in this manual which extract and measure the energy from the sun.

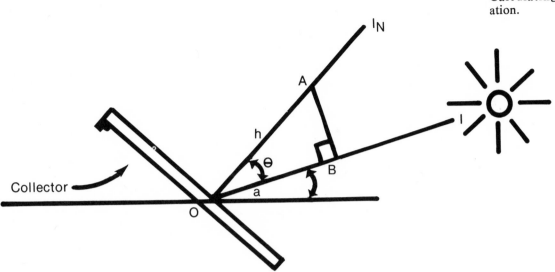

Figure 1—5
Calculating incident radiation.

Geometric analysis

Let the magnitude of the incident radiation I_c striking the collector equal OB when the sun is in line with OB. If the sun were in line with OA or perpendicular to the collector, the angle of incidence θ would be zero and the radiation striking the collector would be at a maximum. With respect to OB, it would be OA.

Therefore, the incident radiation actually striking the collector would be the maximum radiation I_N reduced by the ratio $\dfrac{OB}{OA}$. Or, $I_c = I_N \cdot \dfrac{OB}{OA}$.

In trigonometry, we translate geometrical relationships into the characteristics of angles. With respect to angle θ, OA is the hypotenuse of a right triangle and OB is the side adjacent to the angle θ. The ratio of adjacent side to hypotenuse, or $\dfrac{a}{h}$, is called the cosine of θ. Therefore, $I_c = I_N \times$ cosine of θ. We write this $I_c = I_N \cos \theta$.

Discussion questions

1. The direct or normal rays of the sun form an angle of 90° with a receiving surface. It is these rays that put energy on the surface in greatest concentration, and hence have the greatest heating effect. The angle of incidence of these rays is zero since, by convention, incident angles are measured with respect to a line parallel to the sun's rays and perpendicular to the surface at the point of incidence. From this ex-

periment, can you determine the relationship between the altitude of the sun and the angle of incidence of the sun's rays? Note that as the altitude increases, the angle of incidence decreases, and the heating effect increases.

2. How does the shadow length vary during the day? During the year? Why?

3. How many degrees does the shadow length sweep across the board per hour? Why? Try to measure this.

Determining the Heat Absorbed
by a Closed Automobile

To measure the amount of heat absorbed by the air inside a closed auto left out in the sun.

Objective

When a body absorbs heat and gets hot, its temperature increases. As the temperature of a body increases, the amount of energy it radiates outward also increases. You feel the effects of this when you turn on an incandescent light and the heated filament radiates its heat and light. The temperature of a body will continue to increase until a balance is reached between the rate of incoming energy and the rate of outgoing energy. Under these conditions we say that thermal equilibrium has been achieved. By measuring the temperature increase of the air inside a closed auto left in the sun, you can follow this energy flow quite easily and determine its amount.

Introduction

When the sun's rays strike the auto, a great deal of absorption by the metal exterior occurs. The amount will depend in part upon the color and texture of the surface, with rough black surfaces absorbing much more energy than smooth white ones. Most of the rays that pass through the windows are absorbed by the interior parts. Heating of those parts not directly in line with the sun's rays occurs by conduction, convection, and reradiation from the parts heated directly. Since the air in the auto is fairly transparent to the solar rays, the air is heated to a greater extent by the reradiation from the hot parts than by the direct rays of the sun. This reradiation from the hotter materials consists mostly of the infrared rays, which are much more readily absorbed by the air than the more energetic solar rays.

The heat gained or lost by a substance depends upon three things: the amount of material (mass, or m), the kind of material (specific heat for that material, or s), and the temperature change (ΔT). It is computed from the formula $\Delta H = m \times s \times \Delta T$. The mass of a solid or liquid is easily measured with a platform balance. The mass of an enclosed gas could also be measured, but with greater difficulty. To determine the mass of air that becomes heated in the auto, you will determine the volume of air and compute its mass from its density. Assume a value of 1.08 g/l for the density of air. The specific heat of various materials is tabulated in physics and chemistry handbooks and is quite easy to obtain. For this experiment you may use the specific heat of air as 0.24 cal/g·C°. The temperatures are, of course, measured by a thermometer and the temperature change is computed. If you have a Fahrenheit thermometer, a conversion scale to Celsius temperature is given in the Appendix.

9

Experiment 2

Materials   Two thermometers, meter stick, auto, 10-×-10 graph paper.

Procedure   1. Determine the volume of air in the automobile. This will have to be approximated, but you can make simplifying assumptions and with the meter stick obtain approximate dimensions. Perhaps a station wagon would be good to start with, since its compartment is most regular in its dimensions. Assume that the air space is a regular rectangular solid. Measure the length, width, and height of the compartment with the meter stick.

Figure 2—1

Passenger compartment volume for measuring the energy absorbed by the air inside.

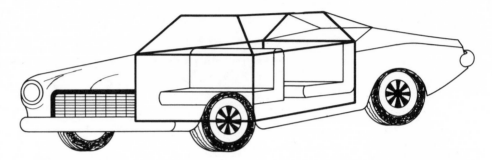

2. Compute this volume from your data by multiplying length × width × height. $V = l \times w \times h$.

3. Make additions or subtractions to this volume by adding the space under the dashboard, for example, or subtracting the volume of the seats, or making other projections to refine your data.

4. Measure the initial temperature of the air in the car and expose the car to the sun. Make sure all of the windows are closed to reduce the rate at which heat is lost from the car. It is best to hang the thermometer shielded from the sun inside and not touching any solid parts. Orient it so that it may be read from the outside without your opening a door and disturbing the air inside.

5. Place the second thermometer in the shade so that the outside air temperature can be measured as well.

6. Measure the temperatures every five minutes or so to observe the short-term variations. Record the time that each observation is taken so that you will be able to determine the rate at which heat is gained or lost. Also record the sky and wind conditions so that their effects can be related to your other data.

7. Look for the highest temperature reached, the time of day it was reached, and how long it stayed there.

8. It is instructive to follow the rate at which heat is lost from the air by conduction, convection, and reradiation as the heat input of solar radiation tapers off; do so by continuing your measurements into the later part of the day.

9. Plot the temperature versus the time of day on the graph paper.

1. Determine the mass of air in the car from your volume measurement **Computations** and the density value from the table in Appendix F—for example, 1.08 g/l.

$$D = \frac{m}{V}$$

$$m = D \times V$$

$$= 1.08 \text{ g/l} \times \underline{\hspace{1.5cm}}$$

$$= \underline{\hspace{1.5cm}} \text{ g.}$$

If you measured the volume in m³, change this to liters.

$$1 \text{ m}^3 = 10^3 \text{ l.}$$

Since the mass of gas also depends on temperature and pressure, perhaps you can come up with a better determination than this by using the gas laws. Note that some of the heated air is likely to leak out of the compartment.

2. Compute the maximum change in temperature that occurred by subtracting the initial temperature from the highest temperature reached.

3. Compute the heat required to raise the temperature of this amount of air from

$$\Delta H = m \times s \times \Delta T.$$

You have just determined m and ΔT, and s is the specific heat of air, approximately 0.24 cal/g·C°.

Note that your answer is in calories. You can convert this answer to BTU (British Thermal Unit) by

$$1 \text{ cal} = 3.97 \times 10^{-3} \text{ BTU or } 1 \text{ BTU} = 252 \text{ cal.}$$

To give you a feeling for how much energy this really is, compare this with the amount of heat required to bring one pint of room-temperature water to a boil.

4. Compute the rate at which solar energy was added to the air by dividing that energy by the time it took to reach the highest temperature.

$$\text{Rate} = \frac{\Delta H}{t}.$$

The rate at which energy is gained or lost is called power. By referring to the conversion tables, you can convert your power measurements in calories per hour or calories per minute to watts.

5. Compute the roof area (or the length times width) to find out the

area over which this solar energy was received. This is, of course, another simplifying assumption. Perhaps the whole south side should be used.

6. The rate at which solar energy reaches the earth is frequently expressed in langleys per minute—one langley is equivalent to 1 cal/cm². As an approximation, you may use 1 langley/min as the insolation on the roof on a clear day in middle latitudes, keeping in mind that this value varies widely depending on the angle of the sun's rays and the moisture content of the air. Multiply this value by the roof area and the total time it took to reach your final temperature. This will give you the total number of calories your roof received. You could include a correction factor for the percent of cloud cover. Compare this value with the one you determined that was absorbed by the air in the car (step 3). Also keep in mind that this value does not include the energy that passed through the windows.

7. Compute the efficiency of this energy transfer from

$$\text{efficiency} = \frac{\text{heat absorbed by the air (step 3)}}{\text{heat received by the roof}} \times 100\%.$$

To improve the validity of this efficiency value, you will want to examine more carefully your assumptions. For example, how much glass area should be included in the collection area determination, and what was the actual insolation at the time of the measurements? Perhaps the clear-sky data supplied in the Appendix for various seasons, latitudes, and times of day could give better results. Note also the difficulties introduced by cloud cover and even wind velocity variations.

Results   Note from the determined efficiency that you have not accounted for all the energy that has been received. You can tell that it is very difficult to keep track of all of the energy—some went to heating the interior material of the car, and much of it was reradiated back to the air since the car was at a higher temperature than the surroundings; some energy was conducted and convected away; some was lost by leakage as air pressure increased.

Now that you have done this experiment, you can repeat it for different cars to see how the color of the roof or the size of the car affects the efficiency of this type of solar energy conversion. This also will tell you something about the size of an air conditioning unit that might be required for different autos. Also, compare your results when measurements are taken on calm days and windy days. It would be interesting to observe seasonal differences for the same auto, as well as what happens when the windows are left open. Figure 2–2 shows the results of one student's measurements.

In principle, all calorimetric methods for measuring solar radiation are done in this manner. The radiation is absorbed by a body, and the increase in temperature produced by the absorbed heat is measured. Try other

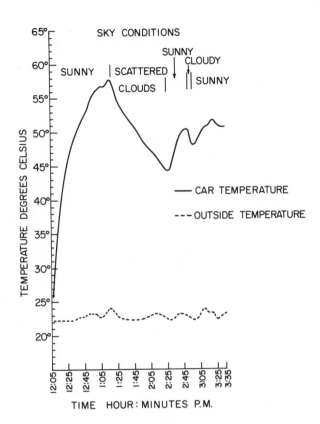

Figure 2—2
Temperature variations inside a closed car. Source: Alex Pidwerbetsky, Linton High School, Schenectady, New York, 1974.

experiments using the thermometer, thermocouple, or thermistor to measure the temperature changes. If you make a solar collector, you will want to measure its energy collection in this same way and determine its efficiency.

Discussion questions

1. What does the graph of the temperature versus time show? Was the rate of temperature increase uniform? Did any clouds cover the sun to reduce the insolation? If so, can this be seen on your graph?

13

EXPERIMENT 3

Determining Heating
Degree-Days per Year

A heating degree-day figure is determined by subtracting from a standard value of 65° F, the mean daily temperature (when below 65° F) —the temperature halfway between the maximum and minimum temperatures for the day. This figure is of particular interest to retail home heating fuel companies, as it is used to help determine the heating demand and hence the frequency for their automatic oil delivery service. It is, of course, important for all energy suppliers also. As a home owner, you will be concerned since your monthly heating bill will be directly related to this figure. The same principles can be applied to the cooling load for summer air conditioning.

It is interesting to follow the number of heating degree-days throughout the season. If you prepare a graph with each day of the month on the x-axis and the heating degree-days on the y-axis, a running record can be kept and the short- and long-term variations can be observed. You can obtain the value for the day by listening to radio or tv weather broadcasts which frequently include the daily high and low temperatures. Perhaps the weather bureau or your local home heating contractor would be willing to give this information to you, although they surely would not appreciate daily phone calls. The best way would be to collect the data yourself with a maximum-minimum thermometer. Regardless of how you obtain the data, subtract the mean of these two values from 65° F and plot it on the graph. For example, a daily mean temperature of 25° F on January 6th equals 40 heating degree-days for that one day.

Keep a running record of the total number of heating degree-days as the season progresses. Compare the results of your work with the regional climatic classification for heating in the United States as shown in table 3–1. Figure 3–1 shows the general location of these regions in the continental United States; the radiation data (mean daily insolation values in langleys/day) refer to the energy received on a horizontal surface.

Regional Climatic Classification for Heating

Table 3—1

Region	Solar Energy Availability	Mean Daily Insolation LY/Day	Heating Requirement	Heating Degree-Days
1	Highest	350-450	Low to Moderate	0-2,500
2	Highest	350-450	Moderate to High	2,500-5,000
3	Highest	350-450	High to Very High	5,000-9,000
4	Moderate	250-350	Low to Moderate	0-2,500
5	Moderate	250-350	Moderate to High	2,500-5,000
6	Moderate	250-350	High to Very High	5,000-9,000
7	Lowest	175-250	Low to Moderate*	0-2,500
8	Lowest	175-250	Moderate to High	2,500-5,000
9	Lowest	175-250	High to Very High	5,000-9,000

Source: Richard Rittleman, *Solar Heating and Cooling for Buildings* Workshop, Washington, D.C., Phase O, Final Report TWR, Report No. 25168.002; NSF/RANN 74-022B.

*This combination does not exist in the United States. Where solar radiation is low it is also likely to be cold. Therefore, the heating requirement is likely to be more than low to moderate.

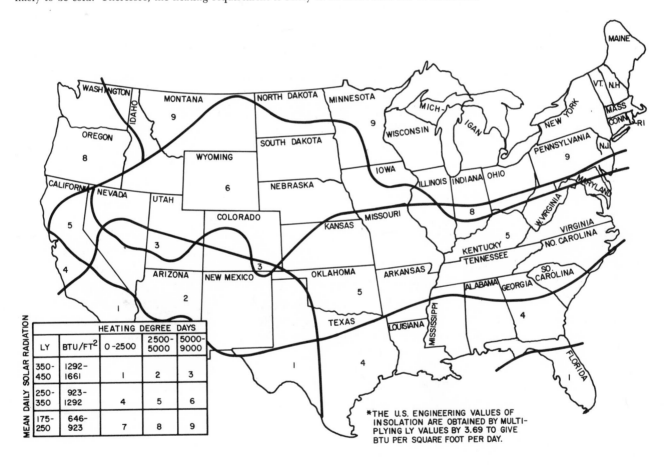

*THE U.S. ENGINEERING VALUES OF INSOLATION ARE OBTAINED BY MULTIPLYING LY VALUES BY 3.69 TO GIVE BTU PER SQUARE FOOT PER DAY.

MEAN DAILY SOLAR RADIATION		HEATING DEGREE DAYS		
LY	BTU/FT²	0-2500	2500-5000	5000-9000
350-450	1292-1661	1	2	3
250-350	923-1292	4	5	6
175-250	646-923	7	8	9

Figure 3—1
Regional climatic classification for the heating season (November-April).

EXPERIMENT 4

Determining the Latent Heat of a Substance

Objective   To measure the effect of incoming solar radiation on the rate of melting ice.

Introduction   The automobile experiment demonstrates how the incoming solar radiation is related to the temperature change of the air in a closed auto. In this experiment, you will observe the effect of solar energy as it causes a phase change from a solid to a liquid without any change in temperature.

Each time a body changes from a solid to a liquid or from a liquid to a gas, energy is required. The ratio of the amount of heat absorbed to the mass of material that changed phase is a physical characteristic for each material and is known as latent heat. For water, the latent heat of fusion L_f is about 80 calories per gram. This shows that for every gram of snow or ice that melts, 80 calories of heat energy were absorbed by the molecules. As a result of this energy, the molecules are now free to move about relative to each other. By measuring the mass of the melted snow or ice, you can easily determine the amount of energy ΔH that was absorbed—from $\Delta H = L_f \times m$. However, if a temperature change occurred either before or after melting, the amount of heat absorbed would have to be included in the calculation from $\Delta H = m \times s \times \Delta T$ as seen in the automobile experiment.

Heat experiments of this type are notoriously inaccurate. You will get the best results if you do this experiment outdoors when the temperature is just 0° C (32° F) and at noon when the rays of the sun have their greatest heating effect. If the temperature is greater than 0° C, you would not know how much of the absorbed heat came from the warmer air rather than from the sun. Should the temperature of the air be less than 0° C, your measurements would not accurately indicate the amount of solar energy required to melt the ice because they would include heat that was lost to the surrounding air.

Materials   Wide-mouth shallow metal can, flat black paint, newspaper and cardboard for insulation, snow or ice cube, clock, centimeter scale, aluminum foil, paper towels, tin snips, file, glass plate, thermometer.

Procedure   1. Make a homemade calorimeter cup with a wide mouth to provide a large energy-collecting bottom surface. A coffee can should work quite well if you cut it so that it is about two inches high. This way, the rays of the noon sun will reach most of the bottom of the can even in the

16

winter. File the top edge smooth to avoid jagged splinters and paint the inside bottom flat black.

2. Wrap newspapers around the outside of your calorimeter cup and then add an outer wrap of aluminum foil. This will reduce the amount of heat exchange to the outside air, and will reduce the amount absorbed from your handling of the cup since your hand temperature is around 37° C.

3. On a day when the sun is shining fairly high in the sky and the air temperature is just 0° C (check this with your thermometer), place a small handful of snow or crushed ice cube in the cup. Try to make sure the snow or ice does not have too much melted water surrounding each crystal. If crushed ice is used, perhaps patting it dry with a paper towel would help.

4. Place the glass plate (make sure it is clean) over the mouth of the cup to reduce the amount of absorbed heat that escapes back to the atmosphere. Set the cup on the ground and orient the cup so that the direct rays of the sun enter the cup's mouth and are absorbed by the black bottom and the ice. Use cardboard and an outer wrap of aluminum foil to insulate the cup from the ground.

5. Record the starting time for the exposure of the ice and cup to the sun. Perform these last three steps as quickly as you can so that the only melting that occurs is due to the heat from the sun.

6. While the ice is melting, agitate the can frequently so that the ice can receive the heat collected by the can as quickly as possible. If you don't, some of the liquid water in the can will begin to warm. This energy will be lost and cannot be included in the measurements. At the same time, keep the mouth of the cup facing the sun, and try not to hold the cup in your hands any longer than necessary. As you can see, this can be a tricky experiment to do well.

7. Just as the last bit of snow or ice has melted, note the time and record it. You will have to observe this very carefully, since it will be difficult to see the last bit of ice.

8. As with all experiments, it is appropriate to do the experiment several times. Your technique should improve, and you will get an idea of how precise your work is by observing how much your results vary at any one time. This is particularly important if you wish to know how the insolation varies between December and February, for example. If this variation is less than that due to the experimental technique or design, you will be unable to observe it.

9. After all the ice is melted, determine the mass of the calorimeter cup and the water it contains by using the platform balance. Pour out the water, dry the cup, and determine the mass of the cup when empty.

10. Measure the diameter of the calorimeter cup in centimeters.

Computations

1. Subtract the mass of the cup alone from the mass of the cup with the water. This will tell you how much snow or ice the sun melted in your calorimeter.

2. Compute the amount of heat the ice absorbed from the sun by

$$\Delta H = L_f \times m$$
$$= 80 \text{ cal/g} \times \underline{\hspace{1cm}} \text{g}$$

where m is the mass of ice you just computed.

3. Determine the number of minutes t it took to melt the ice by subtracting the initial time when the ice was first exposed from the time you recorded when the ice finally disappeared.

4. Determine the area that was receiving the sun's energy. If you oriented the cup properly, this area should be nearly equal to the area of the circle whose diameter equals that of the cup.

$$A = \pi r^2 = \pi \left(\frac{d}{2}\right)^2 = \frac{\pi d^2}{4} 0.785 \ d^2.$$

5. From the last three computations, determine the amount of incoming radiation used per unit area per minute. This will be in calories/cm²·min, or $\frac{\Delta H}{A \cdot t}$.

6. Compare this value with the solar constant of 2.0 cal/cm²·min. You will note that your value is considerably less than this since no more than 43% of this reached the surface, even on a clear day. Also compare your value with the insolation value for your latitude, time of year, and time of day from the data in the Appendix.

Sample calculation

Assume the cup is 10 cm in diameter and the ice cube has a mass of about 50 grams. With a latent heat of about 80 cal/g, the amount of heat required to melt this ice cube is

$$\Delta H = L_f \times m$$
$$= 80 \text{ cal/g} \times 50 \text{ g}$$
$$= 4,000 \text{ cal.}$$

The area of the cup receiving the energy to melt the ice is

$$A = \frac{\pi d^2}{4}$$
$$= \frac{3.14 \times 100 \text{ cm}^2}{4}$$
$$= 78.5 \text{ cm}^2.$$

Thus 4,000 cal/78.5 cm² or 51 cal/cm² of solar radiation is involved. You can see that the ice would take about 51 minutes to melt if the insolation were as much as 1 cal/cm²·min.

The heating effect of the sun depends primarily on the angle that the sun's rays make with the receiving surface, with the direct rays (those making a 90° angle with the surface) being most effective. As a result, the morning or evening sun is less effective than the noon sun, and the winter sun is less effective than the summer sun. In addition, the amount of atmosphere through which the rays pass to reach the surface will influence the heating effect. Greater absorption, scattering, and reflection occur in the morning and afternoon than at noon, since the slanting rays pass through more atmosphere. Therefore, the value of the heating effect you obtained differs considerably from the solar constant. Needless to say, the condition of the atmosphere at the time of your measurement is significant too. Therefore, your experimental results will depend upon the time you do the work as well as your experimental technique.

You can see the need for a less awkward way to obtain solar measurements on a continuous basis if we expect to utilize the sun's energy. Other experiments in this manual will help you do this.

Rather than your waiting for such ideal conditions to perform this experiment (or perhaps while you are waiting), an alternative method suggests itself. You should be able to observe and measure the effect of solar absorption on the rate of phase change by doing two experiments simultaneously—one in the shade and one in the sunshine. In this way, the unmeasured heat transfers to and from the surrounding air will be about the same for both, and the amount of heat absorbed from the sun will be measured by the difference in the time of melting. By removing the unmelted ice from the shaded calorimeter when the exposed calorimeter ice has just finished melting, you will be able to determine the different amounts of heat absorbed in the same time; that difference is attributed to the amount of direct solar energy absorbed. Try it! Perhaps you can also think of ways to increase the amount of incoming radiation to increase the rate of melting.

Another common solar energy effect that utilizes this same principle of latent heat is the solar still. In this case, water in the liquid phase is changed to a gas phase and then condensed back to the liquid phase. Can you devise a way to relate the amount of distilled water collected to the amount of heat required? If you can, the efficiency of this conversion of solar energy can be computed by comparing this value with the insolation at the time of the experiment.

EXPERIMENT 5

Studies with the Solarimeter

Objective

To measure the amount of solar energy reaching the earth and to determine its variability.

Introduction

The automobile and latent heat experiments showed you what can happen when heat is absorbed by a body—a temperature change can occur or a change of state can occur. In this experiment, you will see the effect that solar radiation has on a photocell and will use the effect to measure variations both in full-sky radiation and sun-only radiation. Full-sky radiation includes both the rays from the sun, whether direct or slanting, plus the radiation that has been scattered by the small particles and water droplets in the atmosphere and which reaches the earth from all directions. This scattered radiation might contribute as much as 10% to the total energy received. On a cloudy day, most of the radiation received by your photocell might be due to scattered light.

From your altitude and azimuth studies, you know how the rays of the sun vary with the angle at which they strike the earth's surface. By measuring the electrical output of a photocell and plotting your data, you can relate your results to this variability and to the production of energy. Figure 5–1 shows the effect of this variability throughout the day. In

Figure 5—1

Radiation received on a normal surface and a horizontal surface on 27 February 1974 at Albany, New York.

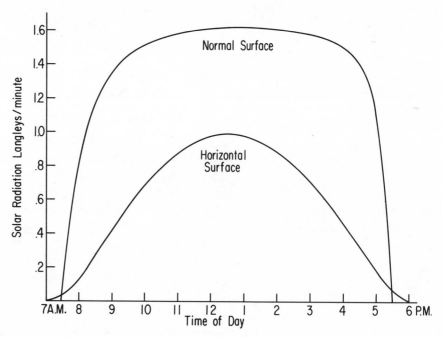

20

addition, by taking complete weather data, you will see that the amount of usable energy from the sun is determined by other factors, such as cloud cover, humidity, atmospheric turbidity, etc. You will be able to measure the turbidity factor with another experiment in this manual.

Materials

Two or more solar or photovoltaic cells (a photographic light meter could be used for some parts of this experiment), milliammeter, watch glass, mounting board and support, silicone cement, resistors, terminal clips, cardboard tube, hookup wire, solder and soldering iron, vacuum tube voltmeter (low range).

Procedure 1

Full-sky radiation measurements

1. Make a solarimeter by mounting one of the photocells and the terminal clips on the board. Solder the cell leads to the terminal clips.

2. Connect the milliammeter to the terminals and observe the response in normal room light. If the needle deflects in the wrong direction, reverse the leads.

3. Here is where you will have to experiment a bit before you will be able to collect your data. As you take your solarimeter out into bright sunlight, observe the meter reading. The idea is to get as large a range of values as possible on the meter—from zero to nearly full-scale deflection—as the solarimeter goes from complete darkness to full, direct sunlight. You will need to match your photocell characteristics to the characteristics of the meter you have. If, for example, full sunlight is too much for your meter, you can try putting a resistor in series between the cell and the meter to limit the current. Try various resistors and arrangements until you find one that will give a good meter response. If your meter is not sensitive enough for your photocell, or if you don't have a milliammeter available, you could use two or more solar cells in series with each other to develop a larger voltage, or in parallel to develop more current. In any case, experiment until you have an array of solar cells which will give you a wide range of current readings for the range of sunlight you have to measure.

4. With a working combination of photocell(s) and meter, you can permanently mount the array on the board and protect it from the weather by cementing a watch glass over it. Make sure the glass is clean. Use silicone cement to make the seal. A groove cut in the board will permit the cell leads to come out from under the glass, which results in a much tighter and neater seal. See figure 5–2.

5. With the apparatus at a location free from obstructions, horizontally rotate the solar cell board to see if any variation in reading occurs as a result of different orientations. If there is a variation, note the horizontal position for a maximum reading, and use this same position for each measurement. Perhaps fixing the apparatus permanently in posi-

21

Figure 5—2
Simple whole sky solarimeter
(pyranometer).

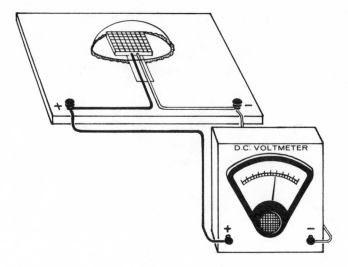

tion on a support stand would be appropriate. (This in itself could become quite a task.)

6. Record hourly readings on clear days, cloudy days, and on the 21st of each month so that you can correlate your data with the altitude-azimuth experiments. Other routine time intervals could be used, according to your own interests.

7. A second set of readings could be taken with the mounting board tilted so that the solar cell surface will receive the rays directly from the sun at an angle of 90° with the surface (angle of incidence=0°). Scattered light will also be received.

8. Make careful records of the atmospheric conditions at each observation (amount of cloud cover, visibility, etc.).

9. Another interesting set of data might be obtained by inverting the board and measuring the response to radiation from the earth. Be sure to keep the height above the ground a constant. Try doing this over different types of surfaces, too.

Procedure 2   Direct sun-only radiation measurements

1. To reduce the amount of scattered light that reaches the photocell array, a cardboard tube could be used. The length of tube to use must be considered. Since the apparent size of the sun in the sky is about ½° (this is known as angular diameter), the solid angle between the center of the cell inside the tube and the open end of the tube must be at least this size in order to receive all of the parallel rays of the sun (see figure 5–3). However, aiming the tube becomes difficult at this small angle. If this angle is too large, too much scattered and diffuse light will enter the tube. Experiment to determine a reasonable and practical tube length that will keep the amount of scattered light to a minimum. Perhaps 10° might be suitable. Professional pyrheliometers have about a 6° aperture. Figure 5–4 shows an alternative method using an oatmeal box.

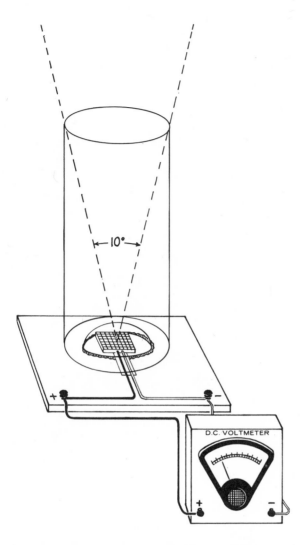

Figure 5—3
Simple pyrheliometer for measuring sun-only or beam radiation.

2. Construct a mount that will support the tube and solarimeter and permit easy aiming. The earth's rotation of 1° every four minutes will require almost continuous shifting of the apparatus. The maximum reading on the meter indicates the correct alignment for data collection. (With appropriate angle-measuring devices, this would be one way to measure the altitude and azimuth of the sun.) Again, this type of mounting can be tricky, requiring your ingenuity. Caution is urged in the aiming of your solarimeter, since the direct rays of the sun can harm the retina of the eye.

3. Record hourly readings on clear days, cloudy days, and on the 21st of each month. Keep careful records of the weather conditions.

Energy measurements

Procedure 3

If you wish to relate the meter readings you obtain to the amount of solar energy received, there are several ways to go about it.

1. One way is to take your data while your instrument is right along-

23

Figure 5—4

Oatmeal box pyrheliometer for measuring beam radiation.

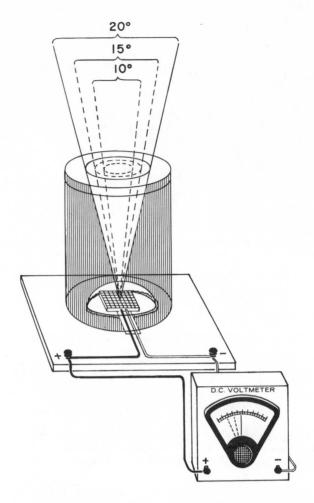

side an instrument that has been calibrated. The New York State Department of Environmental Conservation has an air quality surveillance network throughout the state. Solarimeter data are collected at many weather stations. These data give hourly averages for the particular location. If you live nearby, it would be very worthwhile to visit such a station and calibrate your own instrument directly in this way. Other agencies you might try would include universities, state conservation departments, federal agencies, or private research groups interested in solar energy.

2. If you are not too far away from one of these resources and the atmospheric conditions at your location are not significantly different from those at that agency (this is quite an assumption), you could call or write them for their data on the days for which you have collected data yourself.

3. A third way to get an approximate value of incoming energy is to approach the problem from theoretical considerations. For example: Assume on a clear sunny day that one (1) langley (1 cal/cm²) of radiation is received on a horizontal surface each minute around noon; this would be the time of day that your reading is at a maximum. Then set this maximum reading equal to your unit of radiation intensity— 1 langley. If your meter reading in the afternoon is half of your maximum

reading, then your incident radiation would be 0.5 langley. You will be correct to the extent of which your initial assumption is correct. This procedure of course is not the best, but it does help you to keep in mind what you are trying to do—that is, to measure solar radiation and not just the current flow in a meter.

4. Another consideration of importance is the energy conversion efficiency of your solar cells. Without going into too much technical detail here, you can measure the electrical energy output of your cells fairly easily.

When your milliammeter is reading its maximum current I in full noon sunlight, measure the voltage V across its terminals. A vacuum tube voltmeter works best here because it doesn't draw any electric energy from the solar cells. We will assume that the milliammeter is the electrical device whose energy consumption you wish to know. You might have better luck using a small resistor as the load, however. Power P measures the rate at which energy is consumed, so that Energy E can be expressed as Power × time, or $E=P \times t$.

Since electrical power is $V \times I$,

$$E = V \times I \times t.$$

If V is in volts, I is amperes, and t in seconds, Energy E will be in joules.

$$1 \text{ cal} = 4.18 \text{ joules or } 0.24 \text{ cal} = 1 \text{ joule.}$$

The conversion efficiency can be determined by comparing this power in watts with the power available from the sun, assuming a value of 100 milliwatts/cm² of solar cell area. A better value could be obtained using the tables in the Appendix for your latitude, time of day, and time of year. Compute the energy consumption of the load when the cell is exposed for 1 minute. Here is a sample computation.

Assume the data show as follows:

$$V = 0.51 \text{ volts}, I = 8.5 \text{ ma, and } t = 1 \text{ minute.}$$

Since $E = VIt$,

$$E = 0.51 \times 8.5 \times 10^{-3} \times 60 \text{ sec}$$
$$= 260 \times 10^{-3} \text{ joules}$$
$$= 2.6 \times 10^{-1} \text{ joules.}$$

In calories

$$E = 2.6 \times 10^{-1} \text{ j} \left[\frac{0.24 \text{ cal}}{1 \text{j}} \right] = 0.0625 \text{ cal} = 6.3 \times 10^{-2} \text{ cal.}$$

If you measure the area of your photocell, you can come up with the energy consumption in calories per square centimeter of collector, and compare this value with the assumed average incident radiation of 1 cal/cm².

25

Assume that the cell is $1.3 \text{ cm} \times 0.5 \text{ cm}$.

$$A = l \times w$$
$$= 1.3 \text{ cm} \times 0.5 \text{ cm}$$
$$= 0.65 \text{ cm}^2.$$

The energy output per unit area of collector $=$

$$\frac{6.3 \times 10^{-2} \text{ cal}}{0.65 \text{ cm}^2} \text{ or } 9.7 \times 10^{-2} \text{ cal/cm}^2.$$

The conversion efficiency of the cell compares this value with the insolation value:

$$\text{Efficiency} = \frac{9.7 \times 10^{-2} \text{ cal/cm}^2}{1.0 \text{ cal/cm}^2} \times 100\% = 9.7 \text{ or } 10\%.$$

For technical reasons, the theoretical maximum conversion efficiency for photocells is about 22%. A particularly good solar cell commercially available might be 8% efficient. From your measurements, compute the conversion efficiency of your solar array.

Results

Again, be reminded that the measurements you made tell you the effect the incoming radiation had on your collectors, not the amount of incoming energy.

1. Graph the data in a way that illustrates the short-term and long-term variations in the incoming solar radiation. For example: (*a*) plot actual readings versus time of day; (*b*) plot monthly averages for each hour of the day.

2. If both full-sky and sun-only measurements were made, compare them by plotting each on the same graph.

3. Relate the readings of your milliammeter to the incident solar radiation in cal/cm².

4. If you measured the reflected radiation, determine the net radiation for each hour by subtracting the incident whole-sky radiation and the outgoing radiation measurements made with the solar cell inverted. What does this show? How does this vary during the day?

Discussion questions

1. What are some factors that require consideration in the design of solar energy collectors?

2. Why must routine measurements of this kind be taken over long periods of time?

3. If the conversion efficiency of a solar cell is 5%, where does the rest of the energy go?

Measuring the Solar Radiation Spectrum

To examine some spectral characteristics of solar radiation.

Objective

Introduction

The sun's incoming radiation is absorbed, reflected, and transmitted by the atmosphere as it passes through to the earth. As indicated in classroom activity D (Insolation), not all the wavelengths of the radiation are received on the earth with equal intensity. About 50% of incoming radiation is in the infrared and heat ray part of the spectrum, 41% in the visible part, and 9% in the ultraviolet and X-ray region. This is due to the spectral characteristics of the sun's radiant energy. Absorption of the very short wavelengths (X-rays and ultraviolet rays) by oxygen, nitrogen, and ozone molecules occurs in the outer atmosphere. Except for certain specific wavelengths, the atmosphere is quite transparent to the longer wavelengths. Absorption does take place in certain parts of the infrared region by carbon dioxide, water vapor, water droplets, small particles, and ozone. In this experiment you will attempt to observe separate sections of the spectrum by placing filters in front of the solar cell. Be sure to remember, however, that the data you collect are related to the spectral response of the photocell and filters, as well as the character of the insolation. Success in this experiment will also be dependent upon the quality of the filters that you have available. Figure D–2, page 98, shows the effect of the absorption of solar radiation by the atmosphere.

Materials

Solarimeter of the type made in experiment 3 (sun-only experiment) or other photocell, clear glass filter, variety of colored glass filters.

Procedure

1. Modify the cardboard tube part of the apparatus used in the sun-only solar energy experiments to receive or support the various glass filters in front of the solar cell. Make sure that the filter covers the cell completely.

2. See if you can measure the spectral response of the light transmitted through the filters by using a diffraction-grating spectrometer in your school laboratory. Student models of this instrument are commonly found in the high school. By passing a bright white light through each filter separately and into the spectrometer, see if you can estimate where the center of the band of most intense light falls on the instrument. If it is a direct-reading type, record the central wavelength for each filter. If not, calibrate your spectrometer first, using the wavelengths of known lines, such as helium from a gas discharge tube. An alternative, less interesting method would be to visually compare the color of the filters with a spectral

color chart and read the wavelengths off the chart. Such a chart is frequently seen in high school or college physics texts.

3. Record the maximum reading possible when aiming the solarimeter tube at the sun, using each of the filters available as well as the clear glass and no filter at all. You could also try different combinations to see what variations in readings occur.

4. Take data in the morning, noon, and late afternoon to observe the effects of different amounts of atmosphere on the incoming radiation.

5. Also look for reading variations in atmospheric conditions, such as haze, moisture, clouds, pollution, etc.

Results

1. Record all data neatly in a prepared data table.

2. Plot the readings for each filter versus the time of day.

3. Plot the meter readings versus the wavelength associated with each of the filters.

Discussion questions

1. Are the variations in responses for all filters the same in the morning, noon, and afternoon readings?

2. What conclusion can be made regarding the character of the radiation reaching the photocell through each filter? From this, what can be concluded about the radiation from the sun reaching the apparatus?

Effect of a Converging
Lens on Temperature

CAUTION

Concentrated light rays from the sun can give intensities that are damaging to the eyes. Avoid looking directly at the converging spot on any object under investigation without the protection of welder's goggles or the equivalent.

Objective

To measure the temperature achieved by concentrating the sun's direct rays with a converging lens.

Introduction

The energy received by the earth from the sun has rather fixed characteristics, as you have found out if you have done some of these experiments. The insolation on a horizontal surface might range from about 1.5 langley/minute to nearly zero, depending on the time of day, latitude, and atmospheric conditions. While we cannot change the amount of incoming energy, we can change the effect of the rays on matter. For example, if energy received is 1.0 langley/minute, a converging lens will receive 1 calorie of radiant energy per cm^2 of area during each minute. While a small percent of this energy will be absorbed, most of it will be transmitted and will converge into a region having a much smaller area at the principal focus of the lens. There has been no increase in the amount of energy, yet the intensity has increased. The better the quality of the lens, the more accurately the sun's image can be focused. This concentration of energy increases the temperature of a body placed at this image, perhaps enough to reach the body's melting point or kindling temperature. The temperature achieved also depends on the color of the body. While the total energy involved has not increased, the effect of the energy is more localized, resulting in a higher local temperature.

In this experiment you will try to measure the highest temperature you can achieve with a converging lens.

Materials

Three-inch converging lens (or a Fresnel lens), lens support, assortment of chemical salts with differing melting points, metal bottle caps, chemistry handbook, welder's goggles.

Procedure

1. Look up in the handbook of chemistry the melting points of some of the common chemicals found in your chemistry lab. You are looking for a variety of crystal salts or other materials that cover a wide range of melting points. Make sure they are nonflammable and safe to use for

this purpose. Be sure to check with your chemistry teacher regarding any hazard associated with any you choose.

An alternative procedure is to use commercially available crayons, pellets, or lacquers which have definite melting points. An assortment of these with differing melting points can be purchased for about $2.50 each from many scientific supply companies under the trade names of Tempilstiks, Tempil, and Tempilacq.

2. Place a small amount of each salt chosen into separate bottle caps. An amount that will cover the area of the converging rays is all that is needed.

3. Mount the lens in a way that will permit the rays to converge directly on the salt without wiggling. Place the cap containing the salt with the lowest melting point at the focal point. Again, caution is advised. Do not look directly at the spot without using the welder's goggles.

4. Observe the salt crystals carefully. If any melt, remove the cap and place the cap with the salt of the next highest melting point at the focal point.

5. Repeat this procedure until you find a salt that does not melt. You now know that the temperature achieved by the converging lens is less than this melting point, but higher than that of the previous salt.

6. You could refine this determination by proper selection of other salts whose melting points lie between these two values.

7. By repeating this experiment at different times of day, you can plot the temperature achieved versus the time of day. This graph could then be used to pick a time for experiments that require a particular temperature.

Results

1. It would be interesting to see how small a temperature change you can measure by using this technique.

Effect of a Converging Lens on Energy Collection

Concentrated light rays from the sun can give intensities that are damaging to the eyes. Avoid looking directly at the converging spot on any object under investigation without the protection of welder's goggles or the equivalent.

To measure the energy-concentrating effectiveness of a converging lens using the direct rays of the sun.

As mentioned in the previous converging lens experiment, the total energy involved is not increased by using a converging lens. The area over which the energy is spread is decreased, making the intensity greater. This results in a higher local temperature. Another way to look at the problem is to see the lens as a light-gathering device. If a lens is placed in front of a large body that is illuminated by the sun, the concentrated rays will converge to a small area on the body. The surrounding area of the body will be in the shade of the lens and will no longer receive the energy that has been gathered by the lens. The total amount of heat received by the body will remain the same.

However, if the body receiving the radiation is smaller than the lens, the body will actually receive more energy and the lens can be thought of as a collector. Therefore, this body will receive more energy than it will get without the lens in the same amount of time. In this case, there is an increase in energy received as well as an increase in temperature. Let's try to observe these effects and measure the energy received.

Converging lens and holder, two metal bottle caps, water, flat black paint, newspaper, clock, platform balance, millimeter scale, thermometer.

1. Clean out the inside of both bottle caps and paint them flat black so that they will act as good absorbers of radiation. Set aside to dry.

2. Set up the lens in its support so that one of the caps can be placed at the principal focus of the lens.

3. Measure the mass of each cap when empty.

4. Fill the caps about halfway with water, an equal amount in both.

5. Measure the mass of each cap with water in it.

6. Place one cap in the direct sunlight and one at the principal focus of the lens. Place the caps on some newspaper to insulate them from the ground or support surface. Caution: Don't let the converging rays ignite the paper and start a fire.

7. Record the starting time of exposure and the initial air temperature. You are to compare the time it takes to evaporate all of the water from each cap. This will require a fair amount of vigilance on your part for the evaporation will take some time—at least for the cap without the lens.

8. Record the time each cap runs out of water.

9. Measure the diameter of the lens, diameter of the caps, and diameter of the area of concentrated rays. *Caution: Use welder's goggles to protect your eyes.*

Computations

1. Compute the area of the lens and the area of concentrated rays from

$$A = \pi r^2 = \frac{\pi d^2}{4} = 0.785 \ d^2.$$

2. Determine the collection factor of the lens by comparing these areas.

$$\text{Factor} = \frac{\text{area lens}}{\text{area rays}}.$$

3. Subtract the mass of the empty cap from the mass of the cap plus water, to find the mass of water to be evaporated in each.

4. Determine the length of time required for the water in each cap to evaporate completely.

5. Since the water in each cap required the same amount of heat to completely evaporate it, and the sun supplied energy at a constant rate, then the factor by which the evaporation time differs for each should equal the collection factor of the lens (neglecting heat losses to the surroundings). Compute this time factor.

6. To determine the collection efficiency, the amount of energy used is compared to the amount of energy received. If you assume the intensity of insolation of 1 cal/cm^2·min, you can compute the Energy E collected by the lens from

$E = 1$ cal/cm^2/min $\times$ area of lens $\times$ time to evaporate all the water.

Perhaps a better value of the insolation can be obtained from the data supplied in the Appendix for your latitude, season, and time of day.

7. The amount of heat required to evaporate water can be computed from

$$\Delta H = L_v \times m,$$

32

where L_v, the latent heat of vaporization, is 540 cal/g and m is the mass of water that evaporated from the cap. This neglects any heat gained by the water that increases its temperature. If the water actually boils you know its temperature is at least 100° C, and if you know the temperature at the start of the experiment, you can compute this heat absorption from

$$\Delta H = m \times s \times \Delta T$$

as in the first solar energy measurement experiments. This would amount to less than 15% of the heat required for the phase change in the summertime. Add these two values to measure the total amount of heat required to evaporate both samples.

8. Compare the incoming energy to the lens with that required to evaporate the water. Since some of the energy is absorbed by the lens, the incoming value should be larger than that computed in step 6. The efficiency is given by

$$\text{Efficiency} = \frac{\text{energy output}}{\text{energy input}} \times 100\%.$$

Remember that heat experiments are difficult to do accurately because of so many heat loss variables. It may be that your measurement errors are greater than the amount of energy loss by the lens. Try it anyway. See what happens.

9. You can probably see the relationship between this experiment and the solar stills that you read about. Perhaps you could make a solar still and actually determine its efficiency by measuring the mass of water that was distilled in a given amount of time and computing the energy required from $\Delta H = L_v \times m$. Then compare this figure with the amount of energy received from the sun in that time.

Studies with the Radiometer

Objective To measure the intensity of solar radiation using a vane-type radiometer.

Introduction Many schools have a radiometer which pivots about a vertical axis and consists of four vanes at right angles to each other (see fig. 9–1). One

Figure 9—1
Radiometer solarimeter.

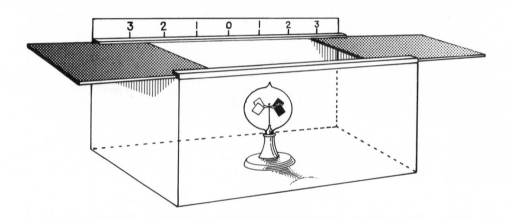

side of each vane is white, the other side flat black. The vanes are mounted in a partially evacuated glass envelope. The black side of each vane absorbs more radiation than the white side, and hence is at a slightly higher temperature. As an oversimplification, when residual gas molecules in the chamber collide with the black vanes, they gain more energy than those striking the white sides. As a result, the molecules exert a larger net force against the black sides when they bounce off. When this force is larger than the static frictional force at the pivot, the vanes begin to rotate. The greater the incident radiation, the greater the speed of rotation.

The vanes rotate too fast in direct sunlight for you to be able to visually use the spin rate as a quantitative measure of the amount of solar radiation. In this experiment, the spin rate is to be gradually decreased to one easily counted, standard rate such as one revolution per second. This could be done in several ways.

1. When two overlapping sheets of polarized material are rotated with

34

respect to each other, the amount of light that passes through both will vary. With one orientation almost none of the light will be transmitted. In this position the two polarizing planes are said to be at 90°. By varying this angle of polarization you can change the amount of light reaching the radiometer, and hence control its speed of rotation. The question then is whether this angle could be related to the amount of full-sky radiation at the time of observation.

2. Another procedure would involve placing the radiometer in a box with a sliding cover as shown in figure 9–1. Can the amount of solar radiation be related to the position of the cover as it controls the amount of entering light?

3. An alternative method would be to place several sheets of tissue paper over the radiometer and measure incident radiation by the number of tissue sheets required to bring the rotation to the standard speed.

Perhaps other possibilities will occur to you as you try this experiment. Be warned, however, that this experiment will require a great deal of patience. The pivot of the four vanes has very little friction, which results in somewhat tedious waiting for the speed of rotation to decrease to your standard value. It is quite easy to observe large changes in speed, but very difficult to observe small changes. Perhaps you would be interested in experimenting to find better ways to measure the speed of rotation and detect smaller changes in speed. In addition, as the amount of light entering the radiometer is decreased, the difficulty in actually seeing the vanes rotate is increased. As the sun sweeps across the sky, the direct rays will enter the apparatus at different angles at different times and will give you another variable to consider.

In spite of these difficulties, or perhaps because of them, you might be interested in trying this experiment. Any results you get will be of great interest, and communication with the author of your successes (or failures) will be welcomed.

Materials

Radiometer, cardboard box large enough to hold the radiometer, two 6-by-12-inch polarized sheets (Edmund Scientific Co. #P–70, 890; $3.00 per pair; Barrington, New Jersey 08007), masking tape, knife, 10-by-10 graph paper, flat black paint, protractor, tissue paper, cardboard.

Procedure

1. Cut off the top of the box and trim the sides so the box just fits when inverted over the radiometer.

2. Cut a 5½-by-8-inch hole in the bottom (now the top) of the box.

3. Paint the inside of the box flat black to reduce reflections.

4. Fasten one sheet of polarized film with masking tape so that the sheet covers the opening. Outline one long edge of the sheet in pencil on the masking tape to make a reference line for angle measurements.

5. With the radiometer in position outdoors, place the box over the device. Make a note of its orientation so that all measurements will be made in the same position.

6. Place the second sheet of polarized film over the first so that both are oriented the same way. Transmission of light through one of them alone is about 30%; with both together, the transmission will vary from 22% to less than 1% when the planes of polarization are crossed.

7. By rotating the top sheet, reduce the amount of light reaching the radiometer so that it will rotate at 1 revolution per second. This can best be determined by seeing that 10 revolutions occur in 10 seconds.

8. Measure with the protractor the angle between the reference line and the long edge of the top sheet. This angle is related to the amount of light that the radiometer receives.

9. Take measurements in the morning, at noon, and in the afternoon to get the daily variations. Repeat the measurements around the 21st of each month to get the annual variations. Be sure to record the current weather conditions as well.

10. If you have a photographic light meter available, you can calibrate your radiometer-solarimeter by taking light meter readings at various angles of polarization when the meter is under the sheets. Plot a calibration curve of meter readings (foot-candles) versus angle between the sheets.

11. If you use tissue paper, record the number of sheets of tissue required to bring the radiometer to its standard speed, and plot this number versus the time of day. If you use a box with a sliding cover, make a scale as shown in the preceding figure and relate the scale reading to the time of day.

Results

1. Plot a graph of the polarizing angle versus the time of day.

2. On the same graph, plot the values obtained for different months.

3. Compare your results with those obtained from the altitude-azimuth experiments and/or solarimeter experiments, and/or the radiation data found in the Appendix.

Measuring Atmospheric Turbidity

1. To measure atmospheric turbidity by means of a sun photometer.

2. To demonstrate that small particles suspended in the atmosphere reduce visibility by increasing turbidity.

The use of the sun photometer is simple and requires little instruction. It can be used to show that visibility can be reduced by pollution on a cloudless day.

The visible portion of the sun's radiation, having wavelengths from about 4,000 to 7,000 angstroms, is called light. When such radiation strikes small particles and droplets in the atmosphere, it is deflected and then strikes other particles and droplets for further deflection. This is called scattering of light, which reduces our ability to see through the atmosphere.

The term atmospheric turbidity is used to describe reduction of the transparency of the atmosphere due to scattering of incoming visible light. We say that visibility is less today than it was yesterday, because increase in turbidity has made it impossible for us to see as far as we could yesterday. This experiment measures how much the turbidity of the atmosphere is increased (visibility reduced) by the scattering of incoming visible light.

Reduction in the brightness of the incoming solar beam of light—by scattering—can be measured by the Volz-type sun photometer. When this instrument is pointed at the sun, light entering a small aperture at the front passes through a filter which transmits a monochromatic beam having a wavelength of about 5,000 angstroms. This portion of the incoming radiation falls on a photocell whose current output, proportional to the brightness of the beam, is indicated by a microammeter.

The theory of operation of the Volz sun photometer is based on the assumption that the logarithm of the ratio of the light intensity sensed by the instrument to the intensity of light at the outer edge of the atmosphere is proportional to the atmospheric turbidity. This relationship is expressed mathematically as follows:

$$\log_{10} J/J_o = -BM,$$

where

J = light intensity measured by the observer.

J_o = light intensity at the outer edge of atmosphere.

B = turbidity coefficient.

M = optical air mass measured by the observer.

To assist in the determination of B, nomographs are available (figures 10–4 and 10–5).

Materials

1. Volz-type sun photometer, such as:
   a. Model 019–1, available from Climet Instruments, Sunnyvale, California, or
   b. Built by students (see next experiment).
2. Nomograph, traced from scales presented herewith.
3. Straightedge.

Procedure

Whether purchased or homemade, the photometer must be calibrated before use to determine its J_o value, which is used with one of the nomographs in determining B, the coefficient of turbidity. If purchased, the instrument will have been calibrated by the manufacturer. If homemade, it will need to be calibrated by you. Even if you use a purchased instrument, calibrating it will be a worthwhile exercise to determine how closely you can duplicate the J_o value supplied with the instrument.

A. Calibration

Choose a completely clear day with a minimum of haze or turbidity (clouds must be absent). Take the instrument to the top of a mountain, the roof of a high building, or other high elevation. Make several observations throughout the day, as described, from mid-morning until mid-afternoon. Avoid obstructions, such as trees or power poles, which would interfere with the path of the sun's rays to the aperture at the front of the instrument. Readings must be taken out-of-doors—never through a window.

Each observation consists of two readings: (1) air mass M, which is determined by measuring the angle of the sun's rays to the horizontal; (2) intensity J of the monochromatic beam striking the photocell (in microamperes).

The front aperture should always be covered with black tape or a lens cover, except when a measurement of light intensity is being taken.

1. Face the sun while holding the instrument case at waist height and level with the horizontal as determined by the bubble in the instrument level.

2. Raise the diopter at the side of the case until the beam of sunlight

shining through the two holes at the front falls exactly on the white spot at the rear of the diopter.

3. Read the value of air mass M from the diopter scale at the point where the scale intersects the top of the instrument case. Record this value as the first reading for Observation #1. (Note: M-readings greater than 4.0 occur only when the sun is very close to the horizon.)

4. Lower the diopter to its original position on the side of the case so that it touches the spot.

5. Tilt the front of the case upward until the beam of sunlight again passes through the holes in the diopter and shines on the white dot at the rear. Remove the aperture cover. Read and record the light intensity J as indicated by the microammeter. See figures 10–1 and 10–2.

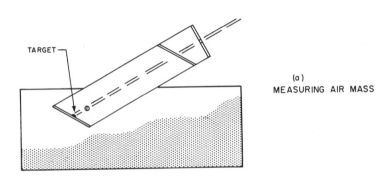

TARGET

(a)
MEASURING AIR MASS

Figure 10—1
Use of diopter to measure air mass.

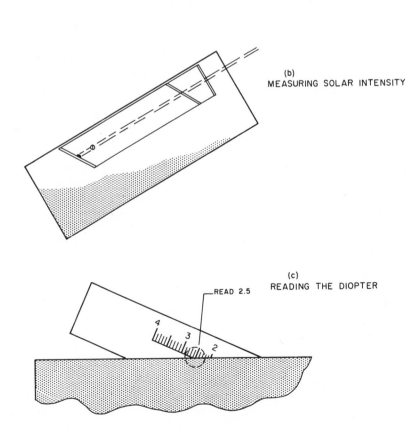

(b)
MEASURING SOLAR INTENSITY

(c)
READING THE DIOPTER

READ 2.5

4   3   2

Figure 10—2
Typical calibration curve.

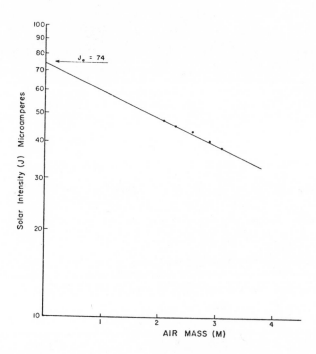

TYPICAL CALIBRATION DATA FOR SUN PHOTOMETER

AIR MASS (M)	INTENSITY (J)
3.1	38
2.9	40
2.6	43
2.3	45
2.1	47

After making observations throughout the day, plot air mass versus intensity values for each observation on semi-log paper, with air mass on the linear abscissa and intensity on the logarithmic ordinate. Fit a straight line through the plotted points and continue it through the ordinate. The intensity at "zero" air mass is the J_o calibration value of the instrument.

B. Determination of turbidity coefficient

1. To determine the turbidity coefficient B at any location, proceed as just described (see "A. Calibration") and obtain readings of air mass M and light intensity J.

2. The turbidity coefficient is read directly from the "B" scale of nomograph I if atmospheric pressure is 29.92 inches $\pm$ 1.5 inches of mercury. If atmospheric pressure is less than 28.42 inches or greater than 31.42 inches

of mercury, use nomograph II and compute the turbidity coefficient *B* from the "A" scale reading. (Note: Making your own nomographs by exactly tracing those supplied with this experiment is recommended if you build your instrument.)

3. Place the intensity scale beside the "$\Delta \log J$" scale of the nomograph, as illustrated, so that the J_o calibration value coincides with the proper seasonal mark near the top of the "$\Delta \log J$" scale.

4. Using the straightedge, connect the intensity (microammeter) reading *J* with the air mass *M* reading for the observation.

5. If nomograph I is used, read the turbidity coefficient *B* at the point where the straight line crosses the "B" scale. If nomograph II is used, correct the "A" scale reading as follows:

$$\text{Turbidity coefficient } B = A - 0.0674 \, P/P_o,$$

where

P = actual absolute atmospheric pressure.

P_o = standard absolute atmospheric pressure.

Note: When using a ratio of absolute pressures, you can use values for any units of measure, provided you are consistent in the numerator and denominator. Standard absolute atmospheric pressure at sea level is 14.7 lb/in², 760 mm Hg, 29.92 inches Hg, or 1013 millibars.

Records

Keep neat and complete records of your calibration and turbidity coefficient determinations.

Problem

Try relating the determined turbidity coefficients to visibility (the estimated miles you can see through the atmosphere). Is it possible to develop a definite correlation factor to translate turbidity coefficients into visibility in miles?

Visibility is defined as "the greatest distance at which a black object of suitable dimensions can be seen and recognized against the horizon sky" under daylight conditions.

References

1. R. A. McCormick, and D. M. Baulch, "The Variation with Height of the Dust Loading over a City as Determined from the Atmospheric Turbidity," APCA Annual Meeting, May 1962.

2. Operating manual, Climet Instruments, Inc., Sunnyvale, California.

3. A. C. Stern, *Air Pollution*, vol. 2, 2nd ed., Academic Press, 1968.

4. Magill, Holden, and Ackley, *Air Pollution Handbook,* McGraw-Hill, 1956.

5. *Handbook of Chemistry and Physics,* 40th ed., Chemical Rubber Publishing Co., 1958–1959.

Source William M. Delaware, Eric M. Holt, and David J. Romano, *Air Pollution Experiments for Junior and Senior High School Science Classes,* Air Pollution Control Association, 1972.

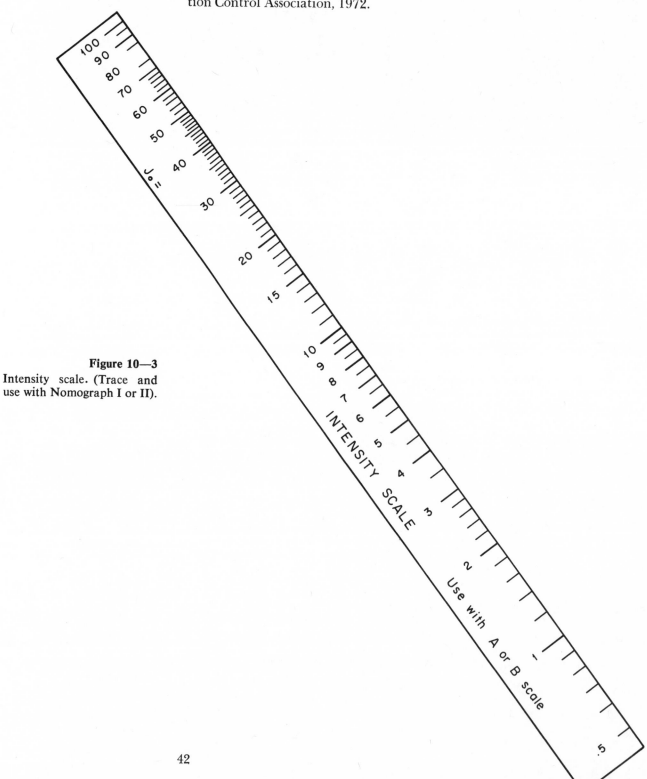

Figure 10—3
Intensity scale. (Trace and use with Nomograph I or II).

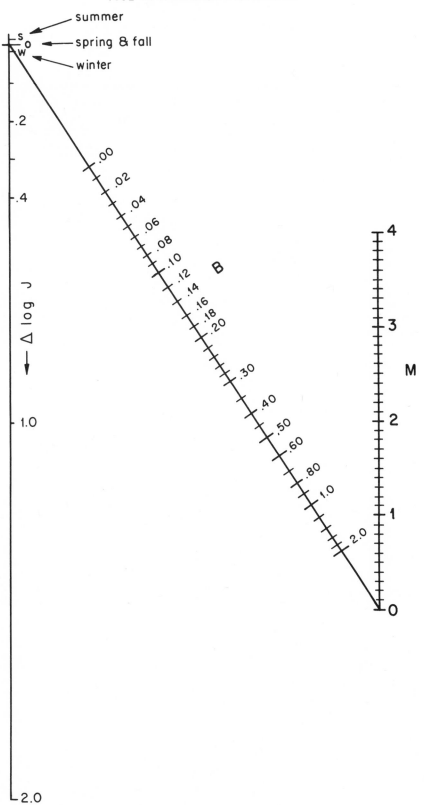

NOMOGRAPH I
B SCALE
(USE WITH INTENSITY (J) SCALE)

Figure 10—4
Nomograph I for B Scale.

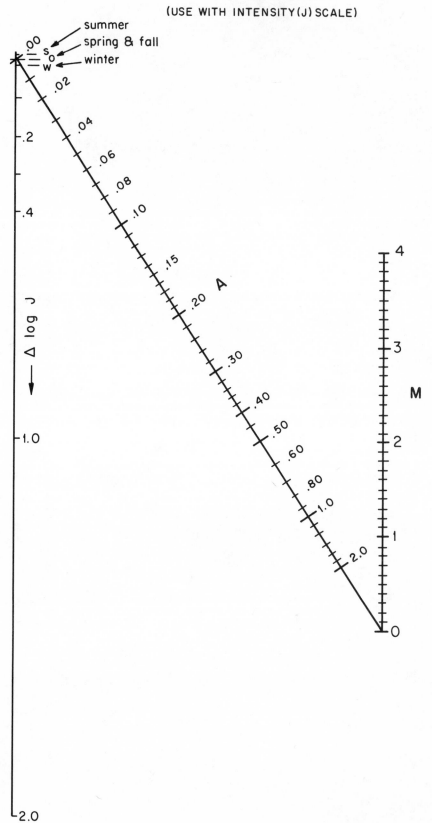

Figure 10—5
Nomograph II for A Scale.

NOMOGRAPH II
A SCALE
(USE WITH INTENSITY(J) SCALE)

summer
spring & fall
winter

Δ log J

A

M

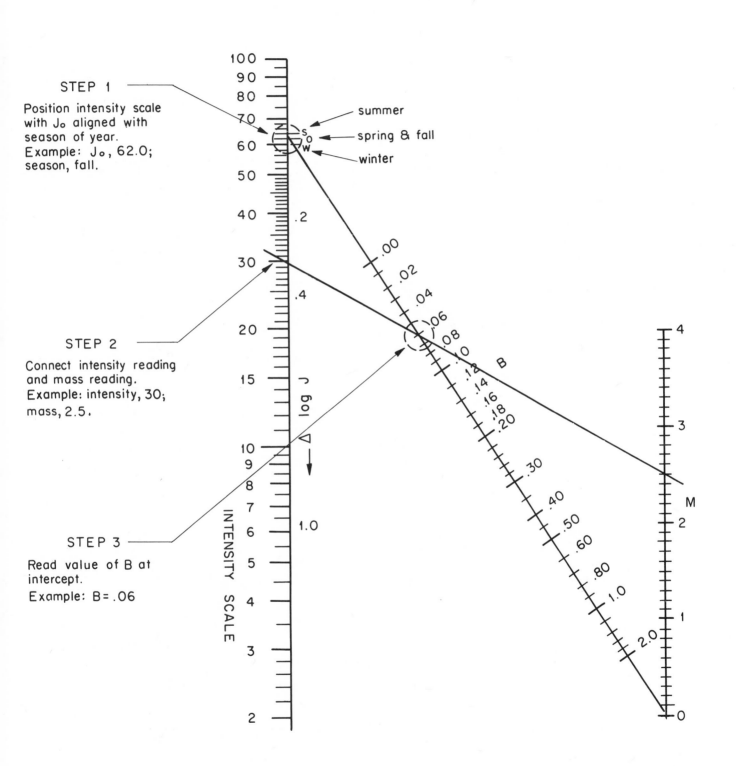

TURBIDITY DIAGRAM
NOMOGRAPH I

Figure 10—6
Using Nomograph I to find
B.

STEP 1

Position intensity scale
with J_o aligned with
season of year.
Example: J_o, 62.0;
season, fall.

summer
spring & fall
winter

STEP 2

Connect intensity reading
and mass reading.
Example: intensity, 30;
mass, 2.5.

STEP 3

Read value of B at
intercept.
Example: B = .06

INTENSITY SCALE

log J

B

M

EXPERIMENT 11

Construction of the Volz Sun Photometer

Objective To construct a Volz sun photometer for determination of the coefficient of atmospheric turbidity.

Construction of the instrument gives a good opportunity to learn how it operates in making measurements of atmospheric turbidity. Also, building an instrument capable of making actual measurements adds to a student's confidence in his or her ability.

Introduction The Volz sun photometer is a relatively simple instrument for measuring the angle of the sun with the horizontal and the intensity of a beam of sunlight, from which a measure of atmospheric turbidity can be determined. It is of simple construction, yet great care must be taken to assure proper dimensions, alignment of parts, and complete sealing of all joints to exclude stray light.

The preceding experiment presented the theory of operation of the instrument. It was seen that the beam of light entering the forward aperture must have a clear straight path through the filter to the photocell, if an accurate measurement of light intensity is to be made. Likewise, the diopter must have its two holes aligned with the white spot at the rear, and all three must form a straight line parallel with the scale itself if an accurate measure of the sun's angle with the horizontal is to be made. The diopter and nomograph scales must be exact reproductions of those illustrated herewith.

Materials

	Item	Approx. Cost	Supplier
1.	Selenium solar cell, output 0.5 v at 0.6 ma, peak spectral response, 5,500 angstroms	$ 1.25	Lafayette Radio Electronics, Model #276–115, or equivalent
2.	Triplett DC Microammeter, Model 120, 0–50 range, 1½" edgewise	23.00	Triplett Corporation, Bluffton, Ohio 45814
3.	Potentiometer, Ohmite Type AB, 5,000 ohms, or equivalent	5.00	Electronic supply store
4.	Kodak Wratten #65 Filter, ½" square	2.00	Photo supply shop

5. Line level	1.00	Hardware store
6. Diopter	5.00	Machine shop fabrication
7. Hardwood, ½′ square, planed if necessary to ¼″ thickness	1.00	Lumber supplier
8. Black paint	Negligible	
9. Small brass wood screws	Negligible	
10. Flat-head wood screw (¼″) and washer	Negligible	
11. Brad, ½″	Negligible	

Total cost approx. $35.00 to $40.00

12. Various tools in wood and machine shops.

Procedure

1. The hardwood instrument case with inner baffle should be constructed according to the drawings in figure 11–1. Outside dimensions of 5½-×-2-×-2 inches are suggested for convenience, but may be modified somewhat if desired. All adjoining edges should be rabbeted and must fit tightly to prevent stray light from entering the case. Available lumber will probably have to be planed to the desired ¼-inch thickness.

2. Before assembly, the sidewalls, top, and bottom should be routed to receive the baffle with a snug fit, as shown on the drawings. The front face of the baffle must be exactly 2 inches from the inside surface of the front end piece.

3. The top should be slotted through as indicated in figure 11–4 so that the level and the microammeter scales can be seen and read easily. Cement the level in place beneath its slot.

4. The front end piece and the baffle should be drilled on center, with a $\frac{3}{16}$-inch diameter aperture in the front piece and a ½-inch diameter hole through the baffle. The centers of both holes must be in alignment when the case is assembled.

5. Install the microammeter in the cover so that its face is visible when the case is assembled. Hold it in place with screws into or through the cover as required. Install the potentiometer at a convenient place in the back piece or in the side behind the baffle.

6. Paint the entire inner surface of the case, including the level, microammeter, and potentiometer, with black paint from a small aerosol can. Be sure to paint the inner surface of both holes. (Allow plenty of ventilation when painting, and do not inhale paint vapors.)

7. Cut a piece of the Wratten filter to fit over the ½-inch hole and fasten it to the front side of the baffle with electrician's black insulating tape.

Figure 11—1
Construction drawing of sun
photometer.

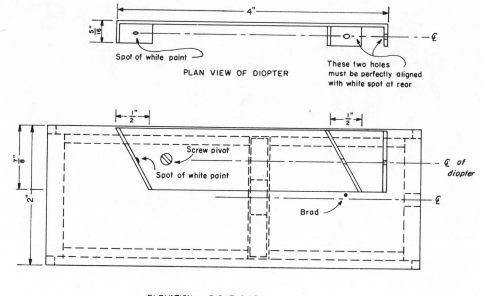

PLAN VIEW OF DIOPTER

Spot of white paint

These two holes
must be perfectly aligned
with white spot at rear

ELEVATION — RIGHT SIDE

SHOWING DIOPTER IN PLACE

Note: All adjoining edges of case to be rabbeted. Sides, top and bottom to be routed
to accept $\frac{1}{4}$" baffle with snug fit.

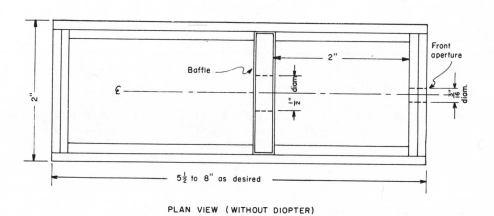

PLAN VIEW (WITHOUT DIOPTER)

8. Center the selenium solar cell over the ½-inch hole on the rear side
of the baffle and fasten it in place similarly. Keep a piece of black tape
over the front aperture whenever the instrument is not in use.

9. Connect the black wire from the solar cell to the microammeter ter-

minal nearest the zero end of the scale. Connect the red wire to the microammeter terminal nearest the 50μa end of the scale.

10. Connect the black and red wires from the potentiometer to the microammeter similarly.

11. Fabricate the diopter from a ¹⁄₃₂-inch brass sheet, or equivalent, as shown in the drawing. Copy the diopter scale exactly from figure 11–2 and attach it to the backside of the diopter (see step 12b, following).

12. Attach the diopter to the right side of the case with a flat-head screw and washer so that (*a.*) its top edge is flush and parallel with the top of the case when the diopter is down against its brad stop, and (*b.*) the scale at 1.0 is flush with the top edge of the case when the diopter scale is vertical or at right angles to the top of the case.

William M. Delaware, Eric M. Holt, and David J. Romano, *Air Pollution Experiments for Junior and Senior High School Science Classes,* Air Pollution Control Association, 1972.

Source

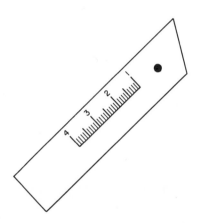

Figure 11—2
Diopter scale.

Figure 11—3
Bottom of case removed, showing baffle, level, and microammeter in place. (Walter Fogg, New York State Dept. of Environmental Conservation.)

Figure 11—4
Diopter raised for reading of air mass (M). (Walter Fogg, New York State Dept. of Environmental Conservation.)

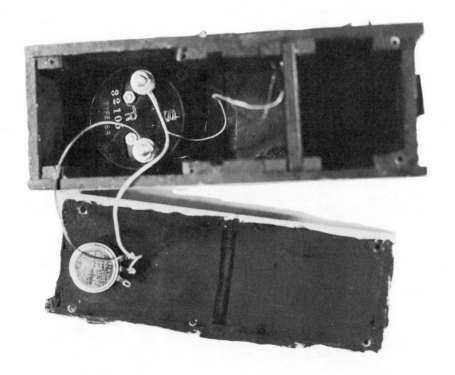

50

AIR MASS (M)	INTENSITY (J)
3.1	38
2.9	40
2.6	43
2.3	45
2.1	47

Figure 11—5
Typical calibration data for sun photometer.

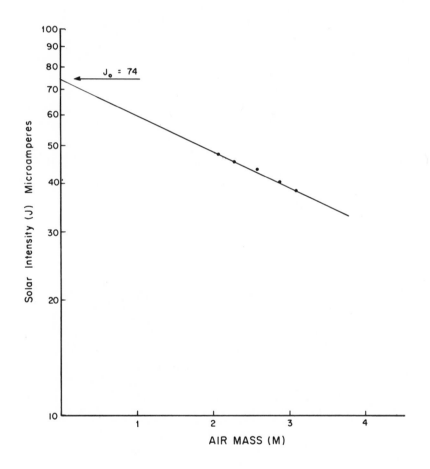

Figure 11—6
Typical calibration curve.

EXPERIMENT 12

Effect of Roof Overhang on the Summer Temperature of Buildings

Objective

To determine the effect that roof overhang has on the air temperature in an enclosed air space.

Introduction

In all energy-consuming systems, whether solar or conventional, care must be taken to see that energy is wisely used and not wasted. This is true for both the designer of the system and the user of the system. All of the energy-saving procedures available to reduce heating fuel consumption in the winter, such as using insulation and storm doors and windows, closing the curtains at night, and others, will also increase home comfort by insulating against the heat of the summer, thereby reducing the energy devoted to air conditioning.

A very simple, automatic way to reduce home energy consumption is through the correct design of roof overhang. The well-planned roof overhang can shade the windows in the summer, and yet allow the winter sun

Figure 12—1
Illustration showing roof overhang and the greenhouse effect.

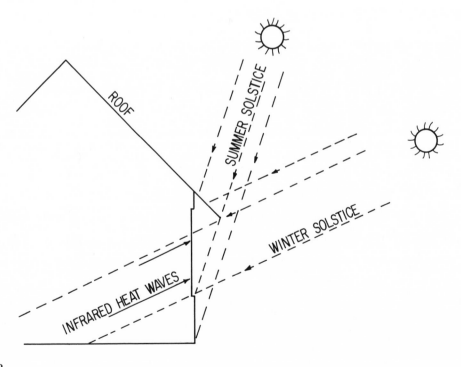

(which is lower in the southern sky) to shine through the windows and warm the furnishings; heating the home in this manner can have a substantial effect, and it reduces the summer cooling load significantly. This experiment will show you the effect such an overhang has on temperature.

Cardboard box about 1-×-1-×-1 foot with a closeable top, window glass about 8-×-10 inches, knife, flat black paint, masking tape, piece of cardboard about 1-×-2 feet, thermometer, metric ruler, 10-×-12 graph paper. **Materials**

1. Cut open one side of the box to make a window, making the opening about one inch smaller on each edge than the glass pane so that the back of the glass will be supported. Also, make the top edge one inch below the top of the box, so that the window is farther from the ground than it is from the roof. **Procedure**

2. Paint the six inside surfaces of the box with flat black paint. Paint one side of the cardboard sheet white to reflect solar radiation striking it; this will be used for the roof.

3. Mount the glass window over the opening with masking tape.

4. Make a hole in the back side of the box about three inches from the top so that a thermometer will just fit.

5. Insert the thermometer so that it will measure the inside temperature and yet be easily read from the outside. The bulb should not be exposed to any direct sunlight that passes through the glass.

6. A weight might be needed to keep the white cardboard sheet in place. Tape a metric ruler along one side of the cardboard so that the amount of roof overhang may be read directly.

Several experiments suggest themselves. Be sure to monitor and record the outside air temperature throughout your experiments.

A. Relate the temperature of the air in the box to the amount of roof overhang for a given sun position.

1. Set the apparatus outside around noon, with the glass side facing south.

2. Determine and record the position of the roof overhang that will cast a full shadow, ¾, ½, ¼, and no shadow on the glass.

3. Pull the roof out so that the overhang completely just shadows the glass and *no* direct sunlight reaches the glass.

4. Watch the reading on the thermometer, and when it has stabilized (perhaps five minutes will do), record the temperature.

5. Move the roof overhang back 1 centimeter so that some direct rays enter the box and become absorbed. Let the temperature stabilize, and record the new reading.

53

6. Repeat this last step, reducing the overhang by 1 centimeter each time and recording the resulting temperature until no overhang exists.

7. Measure the height of the roof above the bottom level of the window. The sun angle measured from the horizontal can be determined from this and from the distance of the overhang when it is just enough to shade the glass completely from the sun. You will note that this angle will vary throughout the experiment, particularly if it lasts several hours, and the variations will affect your results. The angle changes throughout the day due to the earth's rotation and throughout the year as the earth orbits around the sun. It will be instructive for you to keep track of this angle each noon as you repeat this experiment for the different seasons.

8. Plot the temperature versus the amount of roof overhang.

9. Repeat this for different times of the year.

B. Determine the temperature variation in the box throughout the day for a given roof setting.

1. Place the apparatus outside in the morning and have the glass window facing due south. Pick a day that has no clouds, if possible.

2. At first use a roof overhang that will just keep the glass in the shade at noon (local time).

3. Record the temperature in the box every fifteen minutes or so throughout the day.

4. Plot the temperature versus the time of day on the graph paper.

5. Repeat the procedure on successive days with the roof overhang covering ¾ of the glass, ½, ¼, and none of the glass with its noon shadow.

Results   From your graphs, your latitude, and the climatic conditions of your area, what amount of roof overhang seems appropriate for optimum energy conservation?

Energy Consumed by
a Hot Bath

To measure the amount of heat energy used in a hot bath. **Objective**

Any energy-consuming devices which depend on heating effects, such as **Introduction**
incandescent lamps, toasters, irons, dryers, tube-type electronic devices, hot
water and radiant heaters, consume relatively large amounts of electric
energy as compared with other devices such as small motors, fluorescent
lamps, and solid state equipment. While these latter devices do give off
wasted heat, they do not depend upon heating effects for their utility. Find-
ing alternative energy sources, particularly for the heating devices, is an
important step in reducing the demand on our finite fossil fuel supply.
Keep in mind that for every one unit of energy consumed by an electric
heater, three units of fossil fuel energy were required to provide that
electricity. (This assumes that a fossil fuel electric generating plant is
33% efficient in the conversion of energy.) If we all would use a solar-
augmented hot water heater for our domestic needs, the fossil fuel savings
would be significant, as we will see from this experiment (though not a
solar energy experiment per se).

The heat gained or lost by a body can be measured by the temperature
change it causes. This is computed from $\Delta H = m \times s \times \Delta T$, where m is the
mass, s is the specific heat of the body, and ΔT is the temperature change involved. If the temperature increases, ΔT is positive and ΔH is energy
gained.

When you draw hot water from the heater for a bath or shower, water
coming out of the spigot is replaced by water going into the hot water
heater from your cold water supply. This is the water that must be heated
when you take your bath so that hot water will be available for future use.
The more hot water you use, the more energy is required to heat the water
that replaces it. In addition, the water pipes will themselves absorb heat
as hot water flows through, and will lose heat to their surroundings. In-
frequently used hot water pipes and the water they contain lose heat until
they reach the same temperature as their surroundings. In this experiment
you will measure both the energy required for your bath and that absorbed
and lost by the pipes themselves.

Thermometer, ½-gallon empty milk carton, clock. **Materials**

   1. To find the mass of water you use, determine the rate of flow of hot **Procedure**

water and measure how long the hot water is turned on. First, measure the amount of time it takes for the ½-gallon carton to be filled when you turn on the hot water. (Don't burn yourself.) Do this several times to obtain a good average value.

2. Divide ½ gallon by the time you measured, to determine the flow rate in gallons/second.

3. Now draw the bath water, recording the time you start and the time you turn it off. Use only water from the hot water tap.

4. Subtract these two times to find the number of seconds the water was flowing. Multiply these seconds by the flow rate you measured, to find the total amount of hot water used.

5. Measure the temperature of the hot water as it comes from the faucet. Either a Fahrenheit or Celsius thermometer will do, as long as you do not exceed the temperature range of the thermometer.

6. Add sufficient cold water to the tub to bring the water to a comfortable temperature.

7. After the cold water has been running, measure its temperature as it comes out of the faucet. This is close to the temperature of the cold water entering the hot water heater.

8. Now enjoy your bath. While you are doing so, why not measure its temperature for your own interest?

9. The difference between the hot and cold water temperature is the ΔT used to compute the amount of heat required. The thermostat setting of the heater would be somewhat greater than this since some heat is lost from the pipes. The specific heat s for water is 1.0 cal/g·C° or 1.0 BTU/lb·F°.

10. To find the mass of the hot water pipe, you must measure its length from the tub spout to where it emerges from the heater. You will have to approximate the lengths of some segments of pipe that are enclosed in the wall or floor. You can measure the length of the exposed sections.

11. If it is ½-inch copper pipe, assume it has a mass/length of 0.34 lb/ft. If ¾-inch galvanized pipe, assume it has a mass/length of 1.1 lb/ft. Multiply the length of pipe by the approximate value to determine the mass of your hot water pipe. To convert this to grams, use 1 lb=454 g.

12. Find the mass m of water left in the pipe from density $D=\dfrac{m}{V}$ or $m=D \times V$. The volume of water equals the area of the cross section of the pipe $(A=\pi r^2$ or $0.785d^2$—use the inside diameter) times the length of pipe. The density of water is 62.4 lb/ft³ or 1.0 g/cm³.

Computations

1. Galvanized pipe has a specific heat of 0.11 cal/g·C° or 0.11 BTU/lb·F°, while copper pipe has a specific heat of 0.09 cal/g·C° or 0.09 BTU/lb·F°. The pipe will have increased its temperature from about room tempera-

ture to the same temperature as the hot water. Compute the total heat required for your bath from the data you collected: total heat used=heat absorbed by water in the tub+heat absorbed by the pipe+heat absorbed by water remaining in the pipe.

$$\Delta H = m \times s \times \Delta T \text{ (bath water)} + m \times s \times \Delta T \text{ (pipe)}$$
$$+ m \times s \times \Delta T \text{ (water in pipe)}.$$

This computation does not include the heat lost from the pipes to the surrounding air and walls while the water is running.

2. If your determinations were done in the English system, convert the total heat used in BTUs to calories, using 1 BTU=252 calories.

3. Assuming that 1.0 calorie=1.16×10^{-6} kwh of energy, compute the cost of your hot water if electricity is used and costs 3¢ per kwh.

4. What is the amount of energy required from fossil fuel that is used in generating the electricity if your family has a hot water heater? (Assume 33% efficiency for a fossil fuel generating plant.)

5. Assuming that a solar collector of 2-×-4 feet at 30% efficiency is receiving energy from the sun at a rate of 1 cal/cm²·min (929 cal-/ft²·min), how long would it take this solar collector to give you this hot water?

EXPERIMENT 14
Use of Solar Energy in Heating Water

Objective   To demonstrate that useful heat can be produced from solar energy without production of air, water, or land pollution.

These experiments will help to teach students that the sun is an unlimited source of energy that can substitute for at least a portion of the carbonaceous and nuclear fuels used for the production of heat and power. The experiments also demonstrate that useful heat can be gathered from the sun and utilized without producing pollution.

Introduction   Ever since man first discovered fire, he has been relying on the combustion of carbonaceous fuels (grass, dung, fats, wood, coal, oil, and gas) to warm himself and cook his food. With the invention of the steam engine and later the electric power generator, he has also burned fuels to produce power. More recently nuclear fission has been used for power production. All of these fuels produce undesirable byproducts that can foul the air, water, and land, with subsequent detrimental effects on health and economic well-being. Water power, including the rising and falling ocean tides, is being used with the turbine to produce electric power. The power developed by moving air has been used with the windmill. But the availability of falling water, usable tides, and wind is limited.

Our need for electric power has been expanding at an ever-increasing rate and is expected to double every ten years. Yet, in the scramble to keep up with demands for power to operate our industries and to heat and cool our homes, schools, and offices, we have largely overlooked our greatest source of energy—the sun. This has happened in spite of the fact that people have sought from time to time during past centuries to utilize this source of energy. For example, it is said that Archimedes, around 200 B.C., burned the sails of the Roman fleet at Syracuse by focusing mirrors on them. Throughout the ages we have used solar energy to preserve food by drying, and to produce salt and other chemicals by evaporation of sea water. Today, some people still insist on sun-dried laundry.

Let's proceed with our experiment, first trying very simple and primitive methods of heating water by solar energy and gradually working up to more sophisticated methods and equipment.

58

A. First experiment
   1. Garden hose, 25 to 50 feet long.
   2. Thermometer.
   3. Watch.

B. Second experiment
   1. Two large white enamel dishpans.
   2. Black plastic sheeting, large enough to line one pan.
   3. Two sheets of glass or rigid clear plastic ($\frac{1}{8}$- to $\frac{1}{4}$-inch thick), large enough to completely cover a pan.
   4. Thermometer.
   5. Watch.

C. Third experiment
   1. Waterproof bag of clear plastic such as Mylar, about 3 feet square, with the outside of one surface painted black to absorb heat. Inlet and outlet tubing should be provided at opposite edges of the bag. Bag should be sealed to prevent leakage of water. (Mylar can be sealed to itself with heat applied by a hot iron.)
   2. Small tank for storing water, with inlet at top and outlet at bottom. (A five-gallon jerrycan may be used.) Outside of tank should be covered with a blanket of insulation such as fiberglass to retard cooling of heated water. Tank should have a loose-fitting cover or other vent as a safety measure in case too much heat is developed.
   3. Plastic or copper tubing to convey water from bag to tank and from tank to bag.
   4. Thermometer.
   5. Watch or clock.

D. Fourth experiment
   1. Heat absorber, consisting of looped copper tubing ($\frac{1}{4}$ or $\frac{3}{8}$ inch) in a shallow glass or plastic-covered wood or aluminum box. Box should be about 3 feet square (9 square feet) $\times$ 4 inches deep with bottom and side walls completely insulated. Tubing should be looped back and forth, using a bending tool to avoid flattening tubing at the bends, and should then be soldered over its complete length to a flat sheet of copper. Tubing and both sides of copper sheet should then be painted flat black for maximum heat absorption. Lay the assembly of copper sheet and tubing in box with tubing underneath. Provide fittings on both ends of copper tubing which extend through two walls of box.
   2. Tank is in following procedure of third experiment.
   3. Tubing as in procedure of third experiment for circulating water from tank to absorber and back to tank.
   4. Thermometer.
   5. Watch or clock.

Experiment 14

Procedure   First experiment

Lay the garden hose on the lawn at about 8 A.M. on a clear, warm day. Fill the hose and determine both water temperature at the nozzle and air temperature. Leave the hose in this position all day and determine air and water temperatures every half-hour.

Second experiment

At about 8 A.M., set the two dishpans on the roof of your school where they will receive the full rays of the sun all day. Line one pan with the black plastic sheeting. Fill both pans from a cold water faucet. Determine air and water temperatures. Cover both pans with the glass or plastic. Determine outdoor air temperature and water temperature in each pan every half-hour.

Third experiment

Assemble the apparatus on your school roof in the morning of a clear day. Lay down the heat-absorbing bag with its clear surface upward and tilted toward the sun. Mount the tank to the north and a little higher. Connect the tubing to form a closed loop with the bag and tank, so that water will flow from the bottom of the tank to the bottom of the bag and from the top of the bag to the top of the tank. The density of water becomes less as its temperature rises. Thus, warmer water will rise through the heat absorber and flow to the top of the tank, to be replaced by cooler water from the bottom of the tank. Keep the bag shaded until it is full of water. Add water to the tank until tank, bag, and tubing are full. Record the number of gallons required. Replace the tank cover. Remove the shading material so that the bag is in full sunlight. Determine and record outdoor air temperature and temperature of water near the top and bottom of the tank when you start and every half-hour thereafter.

Fourth experiment

Assemble the apparatus on the roof of your school in the morning of a clear day, with the heat absorber tilted to the sun. Greatest heat absorption will occur when the sun's rays are perpendicular to the glass or plastic cover. Keep the absorber shaded until its tubing is filled with water. Mount the tank to the north and a little higher than the absorber. Connect the tubing to form a closed loop with the absorber and tank, so that water will flow from the bottom of the tank to the bottom of the absorber and from the top of the absorber to the top of the tank. Water circulation will be as described for the third experiment. Fill the tank and all tubing with cold water. Record the number of gallons required. Replace the tank cover. Remove the shading material so that the absorber is in full sunlight. Determine and record the following temperatures when you start and every half-hour thereafter: (1) outdoor air; (2) air above copper sheet in absorber; (3) water near top of tank; and (4) water at bottom of tank.

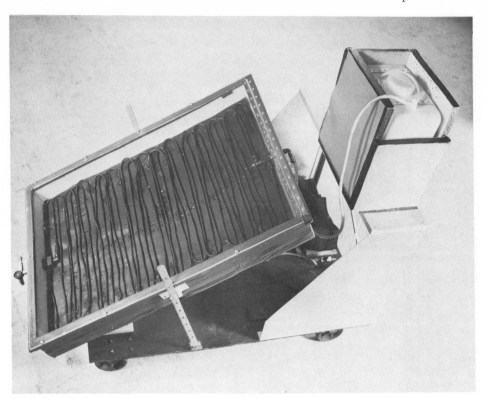

Figure 14—1
Demonstration model of a solar water heater, built and used successfully by New York State Department of Environmental Conservation.

Each experiment should be planned to proceed from about 8:30 A.M. until about 4:00 P.M.; a shorter duration may be possible if water temperature throughout the tank reaches 140° F sooner.

Duration of each experiment

In the third and fourth experiments, water entering the top of the tank may be near the boiling point. Don't touch the tank.

CAUTION

Keep neat and complete records of all of your observations.

Records

1. For each experiment, determine the input in BTU/square foot·hour needed to heat the water to 140° F or to the highest water temperature you were able to achieve. (A BTU, or British Thermal Unit, is the amount of heat required to raise the temperature of one pound of water 1° F. A gallon of water weighs 8.33 pounds.)

Problems

$$\mathrm{BTU/ft^2 \cdot hr} = \frac{\mathrm{lbs\ water} \times (°\mathrm{F\ end} - °\mathrm{F\ start})}{\mathrm{area\ of\ absorber\ surface} \times \mathrm{hours}}.$$

2. If it takes too long to increase water temperature as much as desired, what changes in your apparatus would you propose? Discuss this with your teacher.

3. Can you get enough heat absorption on a cloudy day to increase water temperature?

4. From your own experiments, can you estimate the area of heat-collecting surface and the volume of water storage your family would need for an adequate supply of solar-heated water?

5. Other possible uses for solar energy are suggested for class discussion; for example: solar cooling systems, solar-powered vehicles, and satellite solar-powered generating stations. Supplemental reading of books in your school or state library is suggested.

6. Would it be practical to pipe a solar water heater in series or in parallel with your family's gas, oil, or electric water heater at home and achieve the dual advantage of reducing pollution and cost of heating water? How would you go about it? (CAUTION: Don't start any construction without permission of your teacher and your parents. The authors and editors can't accept responsibility for any plumbing surprises.)

References

1. Peter E. Glaser, "A New Look at the Sun"; *Conservationist,* New York State Department of Environmental Conservation, June-July 1971.

2. D. S. Halacy, Jr., *Fabulous Fireball;* Macmillan Co., 1957.

Source

William M. Delaware, Eric M. Holt, and David J. Romano, *Air Pollution Experiments for Junior and Senior High School Science Classes,* Air Pollution Control Association, 1972. Chapter by Donald C. Hunter. Demonstration model built by John M. Joyce.

15

An Air-Heating
Solar Collector

To measure the collection efficiency of an air-heating solar collector. **Objective**

In order to actually utilize solar energy to reduce the demand for our limited fossil fuels, three primary design considerations are involved: the collection, storage, and extraction of energy from storage for final use. **Introduction**

As heat is conducted, convected, and radiated along each stage, some is lost to the surroundings, thus reducing the efficiency of the system. Therefore, it is important to obtain the highest collection efficiency possible, and then to minimize heat losses. In addition, careful and wise use of the energy that is produced is just as important. To help solve our energy problems, such conservation attitudes will be as important for the users of energy as they are for those who design and manufacture energy-consuming devices. In this experiment, you will make a solar collector using discarded beverage cans to try to increase collection efficiency over what you might realize from a single flat sheet of aluminum.

A solar heat collector has as its purpose the capture and retention of solar radiation as heat until it can be removed for useful work. A dull black-surfaced material of high specific heat, such as copper or aluminum, absorbs radiation as heat better than material of low specific heat. The same material with a light-colored or shiny surface reflects radiation rather than absorbs it. You have probably noticed that dark clothing makes you feel warmer than light-colored clothing. Probably very few of those living in the earth's deserts fully understand the physical laws of solar radiation and heat, but they learned by practical experience centuries ago that light-colored, loose-fitting clothing is more comfortable under the sun's glare; and so today, people in the warmer climates wear voluminous white robes and head coverings.

When you sit near a south window of your home on a sunny winter's day, your shoulder might get quite warm while the window glass is still cold. This is because the glass transmits most of the solar radiation striking it and reflects some back into space, but absorbs very little. On the other hand, the glass effectively blocks in most of the longer wavelength infrared radiation that is directed outward from the heated furnishings of the room.

This is essentially what happens in a solar heat collector. Incoming radiation passes through a glass or suitable plastic cover and is absorbed and retained as heat by blackened metal (usually copper or aluminum). Only a small portion of the heat as long-wave infrared radiation can escape, so we have what is known as the "greenhouse effect." Of course, some heat is lost by conduction through the glass and through insulation on the back side of the collector, and by convection. But by pumping air or a suitable liquid through the collector, we can absorb a large portion of the heat collected and transport it to a heat storage system or to direct use for space or water heating.

The total amount of energy absorbed can be increased by making the collecting area larger. If the roof of a house is to be used as the support for a flat-plate collector, the collecting area is limited. One way we might try to increase the collection efficiency (not the incident radiation), without increasing the roof area, is illustrated by the collector in this experiment (fig. 15–1). Aluminum beverage cans are securely riveted to a flat metal plate in such a way that air may flow past the cans easily. While this increases the effectiveness of the heat transfer to the passing air, it also tends to increase the radiating surface from which heat is lost to the surroundings by radiation and conduction through the cover. Experiment with both collectors to see which gives the best collection efficiency.

Figure 15—1

Hot air collector with and without discarded beverage cans.

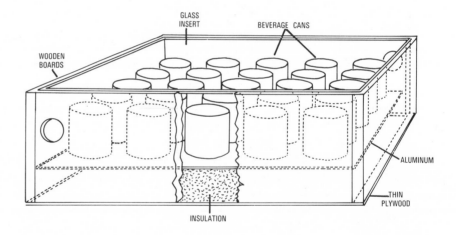

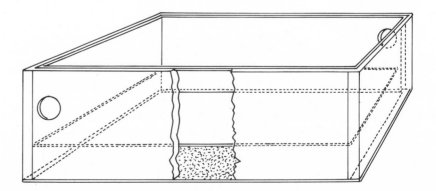

Two pieces of 1-×-1-foot flat aluminum sheet 0.040-inch thick, duct tape, glazing compound, glue, aluminum beverage cans, aluminum rivets, 1-×-3-inch and 1-×-6-inch board as required, electric drill, 13-×-13-inch window glass, flat black paint, nails, band saw, fiberglass insulation 6 inches thick, thermometers (capable of measuring temperatures up to 300° F) .

1. Cut the cans in half with a metal-cutting band saw. You will be able to use both halves. Try to avoid bending the cans.

2. Locate the position on the metal sheet where the center of each can half will be mounted. Have the cans about ½ inch from each other in rows and stagger the rows so the central axis of each can lies between the cans in the neighboring rows. This will tend to make the air flow past more surface area than if the cans were all aligned. Leave a 1-inch border around the edges so that the plate beneath the cans can be mounted on a frame easily.

3. Select the appropriate size drill for the rivets you have, and drill a hole in the center top and bottom of each can half and at each position on the flat plate.

4. Fasten the cans to the plate with rivets, and make sure good contact is made with the plate and the rims of the cans so that good heat transfer will occur by conduction.

5. Construct a square wooden frame 1-foot-×-1-foot-×-3 inches high, using nails and glue. Make provisions for an entrance and an exit port by making 2-inch diameter holes on opposite sides of the frame. Set this on top of a 1-×-6-inch wooden frame holding 6-inch-thick fiberglass insulation.

6. Fasten the plate with the cans in the 3-inch-high frame using small nails. Use glazing compound between the plate and frame to keep the bottom of the collector airtight. This compound could be used in the joints of the wooden frame if necessary.

7. Spray paint the plate (both sides) and cans flat black to make the collector a good absorber of the sun's radiation.

8. Mount the glass plate on the front of the collector using glazing compound between the frame and the glass. Heating duct tape could be used to fasten the glass. Perhaps outside metal corner hold-downs from a hardware store could be used. However you fasten the glass, be sure to protect yourself and others from the sharp edges. The glass will transmit the sunlight readily to the inner black absorbing surfaces, yet retard the flow of infrared radiation outward from the collector. As a result, the air in the collector absorbs the energy and increases in temperature.

9. Wrap 6-inch fiberglass insulation around the four sides of the collector, taping it in place with minimal crushing of the insulation.

10. Build a second collector of the same size, using only a flat sheet of blackened aluminum under the glass. Insulate as before.

Experiments to try

1. With your collector facing the direct rays of the sun at a slope of 45° to the horizontal and with one of the opening ports near the ground, measure the air temperature just inside each of the ports every five minutes to see the effects of the natural convection that become established. Plot temperatures versus the time of day for each collector.

  a. At what time of day is the temperature difference at a maximum?

  b. Compute the heat gained by one collector-volume of air while it passes through the collector at this time in the following way:

  1. Determine the volume of air (ft³) in the collector from $V = l \times w \times h$.

  2. Determine the mass m of air being heated from the density D of air (assumed to be 0.08 lb/ft³ at standard conditions) and the volume V.

  $$D = \frac{m}{V}; \quad m = D \times V.$$

  c. The heat ΔH gained is computed from

  $$\Delta H = m \times s \times \Delta T,$$

  where s is the specific heat of air (.24 BTU/lb·F°) and ΔT, the temperature difference between the entrance and exit ports.

2. Try adding a small blower to the exit port of the collector to establish a circulation of about 1 ft³/min per ft² of glass surface. If you can measure the flow rate of air through the collector and determine the time the air is flowing, you can measure the total volume and hence the mass of air that is heated as it goes through. From this, the total amount of heat for the measured ΔT can be determined. However, this flow measurement is difficult. Perhaps you can develop a way to measure it. One way to avoid measurement difficulty is to assume a reasonable flow rate for both collectors and compute the amount of heat collected on this assumption. Then, the percent of improvement in energy collection can be determined. Since the ΔT changes during the day, you can approximate the total daily heat collection by using the ΔT's at each half-hour and adding the BTUs for each determination throughout the day.

3. Compute the total amount of heat collected during the noon hour (11:30 A.M. to 12:30 P.M. standard time). Compare this with the measured energy of incoming solar radiation for your latitude and season for a south-facing surface tilted at 45°, if you have the data. A gross simplifying assumption would be that this value is about 220 BTU/ft²·hr.

66

Compute the efficiency of your collectors from

$$\text{Efficiency} = \frac{\text{energy output}}{\text{energy input}} \times 100\%.$$

4. Make your measurements at various times of the year so that you can compare the seasonal variations.

5. You could also try your measurements with the collectors at several other fixed angles, such as horizontal, 30°, 60°, and 90°.

6. Now can you devise ways to use the heat? How big do you suppose this collector will have to be to adequately heat your home for one sunny day in winter?

The next logical step is to find a good way to store this energy for later use. One effective answer is an alternative hot-air collector composed of layers of blackened metal window screening. Perhaps you would like to try designing one of your own.

Should you wish to try a hot-water collector, a discarded automobile radiator could be used. Caution is advised here, since boiling temperatures and lack of a circulating pump of sufficient capacity can result in excessive pressure build-up.

EXPERIMENT 16

Solar Energy Storage—Gravel

Objective   To measure the heat storage capacity and the efficiency of a gravel storage unit.

Introduction   We can obtain solar energy to heat our homes, directly, such as through south-facing windows during the daytime only, or we can use a more sophisticated collector and store excess heat for consumption at night or during a few days of stormy weather. We can collect solar heat in water and store it in water for later transfer to room air, or we can collect solar heat in air, store the heat in an inert material such as gravel or broken brick, and then transfer the heat to room air.

This experiment deals with the transfer and storage of solar heat in gravel. Gravel, such as from old creek beds, is inexpensive and convenient to handle. Its frictional resistance to air flow is low since the stones are rounded and smoothed from erosion. Stones from $1\frac{1}{2}$ to 2 inches in diameter have good contact between each other, permitting rapid heat transfer, and the roughly 30% void space allows adequate air passage as the air gives up its heat to the stones and returns to the solar heat collector. The gravel is contained in a wooden bin. Heated air from the solar collector enters the top of the bin, and the air's heat is given up to the stones as it passes through. The cooled air exits from the bottom of the bin and is blown through a duct back to the collector for absorption of more heat. A continuous circulation is maintained as long as the solar collector temperature is higher than the average storage temperature. Thermostats start and stop the fan automatically. Fan or blower size is usually selected to provide air circulation at about $1\frac{1}{2}$ cubic feet per minute for each square foot of solar collector area. Collector, ducts, and storage bin are well insulated and sealed to minimize loss of heat.

In measuring the amount of heat stored, you will be facing one of the more significant problems in solar energy use—namely, the interfacing or joining of the collector and storage units for efficient energy transfer.

Solid rock has a specific heat of 0.205 BTU/lb·F° and a density of 180 lb/ft³. Gravel, with 30% void space, weighs 126 lb/ft³. Considering its overall volume, gravel then has a specific heat of 0.205 BTU/lb·F° × 126 lb/ft³ = 26 BTU/ft³·F°.

Air in the living space of our homes is usually kept at 65° to 70° F. Let's

68

assume that to maintain a 70° F room temperature we need 80° F air, at the lowest, from storage. Thus, any solar heat we add to the gravel in storage must provide a temperature greater than 80° F. If air from the solar heat collector can raise the storage temperature to 120° F, we shall have stored 26 BTU/ft³·F° × (120° − 80°) = 1,040 BTU of heat in each cubic foot of gravel.

Solar radiation on a horizontal surface at Albany, New York, varies from an average of 121 langleys/day (445 BTU/ft²·day) in December, to 540 langleys/day (1,990 BTU/ft²·day) in June. For convenience of construction and for easy snow removal, a solar heat collector as part of a house roof would preferably be sloped at 45° to the horizontal in New York (state) latitudes. Since this sloped surface presents a larger angle of incidence to the sun's rays in winter and a smaller one in summer, about 1,090 BTU-/ft²·day would strike the collector in December and about 1,600 BTU/ft²·day would strike it in June. At a collection and transfer efficiency of 30%, we should be able to deliver an average of 330 BTU on a December day and about 480 BTU on a June day to our gravel storage from each square foot of solar collector.

Assuming we have a *miniature* house requiring about 1,500 BTU/day in December and about 100 BTU/day (mostly for hot water) in June, a solar collector area of 3 square feet and gravel storage of 1 cubic foot is suggested; these figures will ensure good balance and avoid excessive cost when translated to a full-size house. Such a collector/storage combination could squirrel away enough excess heat in September and October to satisfy all heating requirements for the house well into much colder weather. During cold winter weather, solar heat would have to be supplemented with fossil fuel heat as required to maintain the 65° to 70° room comfort.

For a full-size house (having foundation dimensions of about 26-×-50 feet, about 2,000 ft² of heated living space, and a heat loss of around 16,000 BTU/heating degree-day), about 700 ft² of collector and a bin of about 360 ft³ of gravel storage might be used in the vicinity of Albany, New York, to supply about 50% to 70% of the annual heating requirements. By using three storage bins of 360 ft³ each and by operating the bins sequentially in parallel, it should be possible to store more heat during mild weather and reduce fossil fuel requirements still further. Solar heating equipment will add about 15% to 20% to the cost of a conventional house. How soon it pays for itself will depend on the amount of capital borrowed, the interest rate to be paid during the life of the mortgage, and the price of fossil fuel.

For this experiment an air-heating collector will be assumed, but the principles can be applied to any collector you might have made. The dimensions you choose for the gravel bed, the collector size, and the heat losses of your apparatus, will help determine the highest temperature the bed will reach for a given amount of incident solar radiation. If you can come up with a way to actually use the heat stored in your experimental pebble bed, you will have designed and made a complete solar energy heating system. In your design, keep in mind several ideas: (1) transfer the energy from source to use in as few steps as possible; (2) keep the path of heat flow as short as possible; and (3) insulate freely to reduce unwanted heat losses.

Figure 16—1
Gravel storage bin.

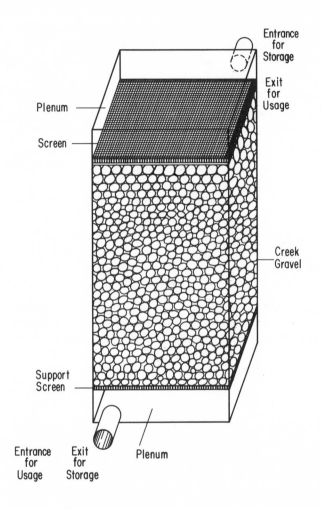

Materials

Thermometers, wood or sheathing material, glue, small electric fan, screened creek gravel (about 1¼ to 2 inches in diameter), solar collector, insulation material, 2-inch diameter duct work (old vacuum cleaner hose should work nicely), bathroom scales, 10-×-10 graph paper.

Procedure

1. Decide on the volume of gravel material that will be suitable for the size of the collector you have, using the considerations mentioned above.

2. Construct a container of the appropriate volume (1 ft³) to hold the pebbles, using plywood or sheathing material. Make sure that all the edges are sealed tightly. The top should also make a tight fit so that the circulating air does not escape from the system.

3. Provide an entrance port on one side near the top and an exit port on the opposite side near the bottom, so that the air flow comes in contact with as many pebbles as possible.

4. Determine the amount of rock material you are using. If a small bin is made, weigh it when empty, and then weigh it when filled with pebbles. Weighing can be avoided by assuming a density of 126 lb/ft^3 and a specific heat of 26 BTU/ft$^3 \cdot$F°.

5. Mount the fan in the exit port of the collector.

6. Connect the duct work from the fan to the entrance port of the storage bed, and from the exit of the storage bed to the entrance port of the collector.

7. Make provisions to insert a thermometer in the entrance and exit ports of both units so that you can tell exactly what is happening in various parts of the system.

8. Move the apparatus into position outside. You might wish to simulate basement storage by shading the storage unit from the sun for your experimental measurements. Keep the collector shaded until you have measured the initial temperatures of the air in the system.

9. Fill the storage bin with the pebbles and secure the top.

10. For your first measurements, you may want to keep the collector surface at right angles, or normal, to the direct rays of the sun for maximum heating effect. Then slope it permanently at 45° facing south. You can compare results later for the normal and 45° collector positions.

11. Measure and record the initial temperatures of the air in the entrance and exit ports of each unit.

12. Turn on the fan and expose the collector to the sun.

13. Measure the air temperature in the system at all four locations every five minutes to start with. Monitor the outside air temperature as well as the weather conditions.

14. Continue the temperature measurements into the late afternoon and evening to find how long it will take for the stored heat in the pebbles to be released to the atmosphere. The pebbles will then be the same temperature as the outside air.

15. Repeat the experiment after you liberally insulate the storage unit with fiberglass or rock wool. Note: any heating effect that might occur due to the electric fan motor getting hot and adding energy to the system, as well as any adiabatic heating that would occur as a result of the fan compressing the air, should be small enough to be negligible.

1. Plot the temperature versus the time of day for each of the locations on a 10-×-10 sheet of graph paper.   **Results**

2. Compute the total amount of heat stored in the storage bed from

$$\Delta H = m \times s \times \Delta T,$$

where m is the mass of rock; the specific heat s of the rock is 0.205 cal/g·C° (0.205 BTU/lb·F°) or 26 BTU/ft³·F°; and ΔT is the difference between the initial temperature and the maximum average temperature reached by the pebbles in the storage unit.

To learn the efficiency of your system, you must determine the amount of energy received by your collector. One approximation would be to assume an average intensity of radiation. Use an appropriate value according to the table, in Appendix D, of daily insolation by months at various latitudes. Using the area of your collector and the time it took for the pebbles to reach their maximum temperature, compute the assumed incident energy. This will be valid within the limits of your experimental errors only to the extent your assumption is correct. If you have a calibrated solarimeter to give you the value of the incident radiation, the results will have greater validity.

From the volume of the air in your collector ($V = l \times w \times h$) and the density of air assumed to be 1.08×10^{-3} g/cm³, the mass of air in the collector at one time can be determined from $D = \dfrac{m}{V}$. A more precise method would be to use the general gas law to find the mass of gas, which takes into account the atmospheric pressure and temperature as well as volume. To determine the total mass of air heated during exposure to the sun, the flow rate must be measured. This can be difficult. One method would be to insert a puff of smoke into the input of the collector. Measure the time it takes for that batch of smoke to replace the air that is in the collector. Try this several times to come up with an average time that the air is replaced. The flow rate $R = \text{volume/time}$. Multiply this rate by the time it took the rocks to reach their maximum temperature, to get the total volume of air that absorbed energy in passing over the collector. Compute the total mass and, from $\Delta H = m \times s \times \Delta T$, the total energy absorbed by the air. Now you can compare this figure with the energy stored in your rock bed, to determine the storage efficiency.

$$\text{Storage efficiency} = \frac{\text{energy stored in pebbles}}{\text{energy absorbed by the air}} \times 100\%.$$

In your energy considerations, do not forget to include the energy required to run the fan, and remember that three times its energy consumption was required as fossil fuel to give you the electricity from a generating station.

Discussion questions

How can you now extract this stored energy and actually use it? What would be the resulting net efficiency, compared with the values you obtained?

72

Solar Energy Storage— Glauber's Salt

As we explained in the latent heat experiment (4), heat is required to change a substance from its solid to its liquid state at its melting point. This heat is called heat of fusion and is absorbed by the liquid. When the substance cools, it ordinarily changes from a liquid to a solid at its melting point and releases the heat of fusion, sometimes called heat of crystallization. At times it is possible for cooling to continue far below the normal crystallization point without having crystallization occur. This is called supercooling. When this happens, the only heat recovered from the substance is that which is released as the temperature of the substance drops.

Introduction

STORING

RADIATING

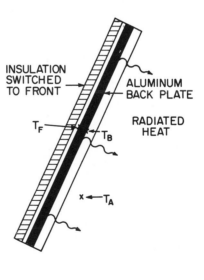

Figure 17—1
"Salt box" module. Source: Keith Dayer, Atmospheric Sciences Research Center, 1974.

Figure 17–1 shows a design one student made to store solar energy using the latent heat principle. The salt that he used was decahydrated sodium sulfate ($Na_2SO_4 \cdot 10H_2O$), also known as Glauber's salt. He monitored the temperatures on the front and back plates of the collector as well as the

Experiment

ambient temperature. The results of these measurements are shown in figure 17–2. The time during which latent heat was absorbed was between 2:30 P.M. and 5:00 P.M., after which a sharp increase in temperature was observed, indicating that all of the salt had melted. You can also see that even though the temperature of the salt was below its melting point after being indoors for awhile (after 6:30 P.M.), the temperature continued to drop. The observation was made that the material was still mostly a liquid. This was a supercooled liquid and its latent heat of fusion was not released and not available as a heat source. This behavior of the material is not predictable; it will not always change phase at 32° C (90° F) during the cooling process. If it did, it would make a fine heat sink for solar energy heating because of its appropriate melting point and comparatively low cost. Research is being done on the addition of nucleating agents to help trigger recrystallization at the melting point.

Figure 17—2
Heat storage module. Source: Keith Dayer, Atmospheric Sciences Research Center, 1974.

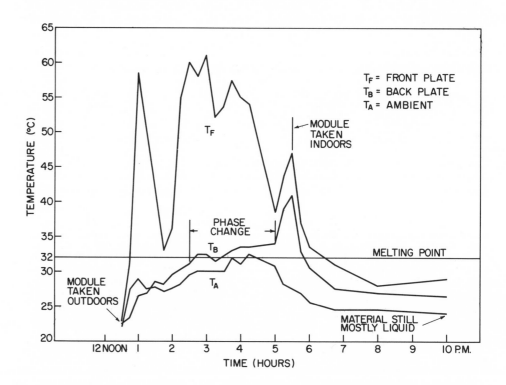

Perhaps you would enjoy experimenting along these lines. Study the properties of Glauber's salt and try to get it to give up its latent heat of fusion regularly and predictably. Perhaps you could find other suitable salts. Should you succeed, you would have to determine how to actually utilize the heat that is released as the salt box cools. One obvious use would be as a space heater in the cool of the evening, in a similar way that hot water bottles or bed warmers were once used. Other aspects would have to be studied, such as how long the salt box can be used without any maintenance and how long it can withstand severe weather. You should also be concerned about the different rates of expansion of the salt and the container, particularly during the phase change. How can this be taken care

74

of in the design of the collector? Another problem is the stratification that sometimes occurs in the fluid, which results in a nonuniform medium. (Try a maximum depth of 1/4 inch.) As you can see, there is still a lot that can be learned by attempting to apply scientific principles to make working devices.

The advantage that such storage has over water storage can be illustrated by the following problem.

An average household might have a need to store about 3/4 of a million BTUs of solar energy to help heat the home. Compare the space requirements for the storage of this heat using Glauber's salt with those required if water were used as the storage medium. Assume that (a) the heat is stored in water, taking it from 80° F to 120° F; (b) the heat is stored in Glauber's salt for the same temperature range.

1. For water

   Heat gained by water in warming from 80° F to 120° F is:

   $$\Delta H = m \cdot s \cdot \Delta T$$

   $$m = \frac{\Delta H}{s \cdot \Delta T}$$

   $$= \frac{3/4 \times 10^6 \text{ BTU}}{1 \text{ BTU/lb} \cdot \text{F}° \times 40 \text{ F}°}$$

   $$= 1.88 \times 10^4 \text{ lb}$$

   $$= 9.4 \text{ tons.}$$

   Since density D is mass m divided by volume V,

   $$D = \frac{m}{V}$$

   $$V = \frac{m}{D}$$

   $$= \frac{1.88 \times 10^4 \text{ lb}}{62.4 \text{ lb/ft}^3}$$

   $$= 3.0 \times 10^2 \text{ ft}^3$$

   $$= 2,244 \text{ gallons.}$$

2. For Glauber's salt

   Specific heat of solid $s_s = 0.46$ BTU/lb·F°.

   Specific heat of liquid $s_l = 0.68$ BTU/lb·F°.

   Density = 91.3 lb/ft³.

   Melting point = 90° F.

   Latent heat of fusion $L_f = 104$ BTU/lb.

The total heat gained by Glauber's salt from 80° F to 120° F includes heat gained as the solid warms plus the heat gained as the solid melts plus the heat gained as the liquid warms.

$$\Delta H = m \cdot s_s \cdot \Delta T + m \cdot L_f + m \cdot s_l \cdot \Delta T$$
$$= m \ (s_s \cdot \Delta T + L_f + s_l \cdot \Delta T)$$
$$m = \frac{\Delta H}{s_s \Delta T + L_f + s_l \Delta T}$$
$$= \frac{0.75 \times 10^6 \ \text{BTU}}{0.46 \ \text{BTU}/\text{lb} \cdot \text{F}° \times 10\text{F}° + 104 \ \text{BTU}/\text{lb} + 0.68 \ \text{BTU}/\text{lb} \cdot \text{F}° \times 30\text{F}°}$$
$$= 5.8 \times 10^3 \ \text{lb}$$
$$= 2.9 \ \text{tons}.$$

$$\text{Since } D = \frac{m}{V},$$
$$V = \frac{m}{D}$$
$$= \frac{5.8 \times 10^3 \ \text{lb}}{91.3 \ \text{lb}/\text{ft}^3}$$
$$= 64 \ \text{ft}^3.$$

You can see that the water heat storage requires almost five times the amount of space as the Glauber's salt heat storage. This space savings would result in reduced costs for insulation and construction.

Thus, overcoming the tendency of Glauber's salt to supercool is a useful goal for the experimenter. A good summary of design considerations and experimental results with this and other eutectic salts can be found in *Storage of Solar Heating/Cooling* by Maria Telkes of the University of Delaware (published by the Institute of Energy Conversion, University of Delaware, Newark, Delaware). Table 17–1 lists several inexpensive salt hydrates whose properties are currently under investigation. Perhaps you would like to work with some of these.

Another heat storage material you may wish to study and compare with Glauber's salt—and one that is not subject to supercooling—is a mixture of stearic acid (a wax) with oleic acid (a liquid).* Use about five parts of stearic with one part of oleic. The molecules of these substances are quite similar in composition except that the oleic acid contains a double bond. This mixture has a crystallization (and melting) temperature of about 50° C.

---

*Refer to expired U.S. Patent #2,726,211, General Electric Co., issued to V. J. Schaefer, 1955.

SOME INEXPENSIVE SALT HYDRATES

Table 17—1

	Name	Chemical Compound	Melting Point (°F)	Heat of Fusion BTU/lb	Density lb/ft³
1.	Sodium thiosulfate pentahydrate	$Na_2S_2O_3 \cdot 5H_2O$	118–120	90	104
2.	Sodium sulfate decahydrate	$Na_2SO_4 \cdot 10H_2O$	88–90	108	97
3.	Calcium nitrate tetrahydrate	$Ca(NO_3)_2 \cdot 4H_2O$	102–108	60	114
4.	Disodium phosphate dodecahydrate	$Na_2HPO_4 \cdot 12H_2O$	97	114	95
5.	Sodium carbonate decahydrate	$Na_2CO_3 \cdot 10H_2O$	90–97	106	90
6.	Calcium chloride hexahydrate	$CaCl_2 \cdot 6H_2O$	84–102	75	102

EXPERIMENT 18

Solar Energy Storage—
Electrochemical

Objective   To measure the overall efficiency of a solar electrical energy collector and storage system.

Introduction   The variable nature of solar energy necessitates some sort of storage of the collected energy to meet the needs as required. You might like to add a bit of chemistry to your experimentation by storing in batteries the electrical energy generated by solar cells. It is quite easy to make a simple chemical storage cell consisting of two lead plates as electrodes and sulfuric acid as the electrolyte. However, particular attention has to be paid to good laboratory procedures and techniques under the supervision of your science teacher; the experiment involves handling hazardous materials, use of the hood, rubber gloves, and proper disposal of spent materials. An alternate technique would be to use existing rechargeable batteries of the hearing-aid or transistor-radio variety to store the energy of a small solar cell array. You could also reclaim some old auto batteries, many of which have used less than 50% of their plate material.

Whichever battery you use, it is important to measure the amount of solar energy coming into the solar cell, the unit's energy output to the storage input, and then the storage output to an energy-consuming device such as a small lamp or a motor doing work that can be easily measured. Two important considerations which determine whether such a complete energy system will actually be useful are the overall efficiency of the whole system and its cost per unit of useful energy delivered.

Materials   Battery jar, lead plates, connecting wire, milliammeter, voltmeter, nylon or fiberglass screening, concentrated sulfuric acid, low-voltage DC power supply, ammeter, hydrometer, solar cells sufficient in number to give up to 2.7 volts at 0.5 to 1.0 amperes.

Procedure   Two thin 10-by-10-centimeter lead plates can be used to form the electrodes for your storage cell. Place them in a battery jar or beaker as close together as possible, using a thin fiberglass netting separator to keep them from touching. Porous polypropylene or polyvinylchloride may be used; nylon, however, is not stable in H_2SO_4. For an electrolyte, use a 9-10 normal solution of about 50% acid and 50% water so that the specific gravity is 1.4. Apply a 2- to 3-volt charging potential to the plates, using the power supply. The plates must be very porous to increase the reacting surface area and the storage capacity of the cell. This can be done by reversing the charging polarity. This is accomplished by applying a 6-volt DC potential across the plates. The anode (+) plate will oxidize to lead oxide and the cathode (−) will remain lead with hydrogen gas being formed. (CAUTION: the gas is highly flammable; do not let it accumulate.) By reversing the charging direction many times, the roughening up of both electrodes can increase the surface area to perhaps 100 times the geometric area of the plates and hence increase the energy storage capability of your cell. Without 100 or so of these charge reversals, of 5 to 10 minutes each, the plates will be unable to maintain the charge. As an alternate to this procedure, you could use conveniently available lead acid or nickel cadmium cells, or scavenge the plates from an automobile battery. Monitor the voltage and current conditions as you prepare the cell so that you can follow what is going on inside.

The chemical reaction for the operating battery is as follows:

$$2PbSO_4 + 2H_2O \underset{\text{discharging}}{\overset{\text{charging}}{\rightleftharpoons}} \overset{\text{+electrode}}{PbO_2} + 2H_2SO_4 + \overset{\text{−electrode}}{Pb.}$$

Specific gravity of acid = 1.28 to 1.10 upon discharging,
7.14 normal solution,
37.5% by weight.

During charging, the reactions are as follows: At the positive electrode

$Pb^{+2}SO_4 \rightarrow Pb^{+4}O_2 + 2e$ (charge reaction-oxidation/anodic
discharge reaction-reduction/cathodic).

At the negative electrode

$Pb^{+2}SO_4 + 2e \rightarrow Pb^0$ (charge reaction-reduction/cathodic
discharge reaction-oxidation/anodic).

During discharging, the reactions are such that the cathode becomes the anodic electrode in that it gives up electrons, but in this case to the external circuit. Thus its polarity remains (−).

Figure 18-1 summarizes the results of the chemical reactions in the cell in terms of the potential across the terminals of the cell. Try to measure the actual charging and discharging characteristics of your storage cell by monitoring the voltage, current, and time.

If you try to charge your storage cell with the output of a solar cell, be sure to put a rectifier of some sort (crystal diode or silicon rectifier) in the circuit. Observe the correct polarity so that it will permit current to flow from the solar cell to the battery when the sun shines on the cell, and yet prevent the opposite flow so the battery will not discharge itself through the solar cell when taken inside.

Should you wish to experiment using rechargeable batteries, a typical charging-discharging voltage curve for a small nickel cadmium cell is shown in figure 18-2.

You can determine the efficiency of your collection system by measuring the energy input to the solar cells or wind generator and measuring the total output from the storage cell. Find the total energy received by your solar cells by using their area, the duration of their exposure to the sun, and the radiation data for the time of day, time of year, and your latitude. Alternately, use radiation data available from some local weather stations or Environmental Conservation Department monitoring stations. Try using the solar cells both in the horizontal position and directly facing the sun. For measuring the energy output, try running a small electric motor from the storage cell and measure the amount of work it can do in lifting a weight, realizing of course that the efficiency of the motor will have to be included in the overall efficiency. To determine the motor's energy input, monitor the voltage and current output to the motor and the time.

An alternate method for measuring the energy output of your cell is to do a standard calorimeter experiment measuring the amount of heat absorbed by water when you connect your cell to a small heating coil of wire. Most standard physics texts show you how to use this technique.

Another factor you should consider (and measure) is the effect on the efficiency of the system as you make your collection area and storage units larger. If you double the size of each, will you double the amount of energy you can generate? Try it.

As you might guess, this experiment could very well require your whole three or four years in high school or college to complete. It will give you a real taste of true scientific research. You will make many mistakes as you try to make things work, but that's part of the learning process. Keep at it! Talk over your work with others who might help you overcome difficulties and keep you from getting discouraged. Become aware of alternate ways to accomplish your objectives and other avenues for additional investigations. Most importantly, have fun.

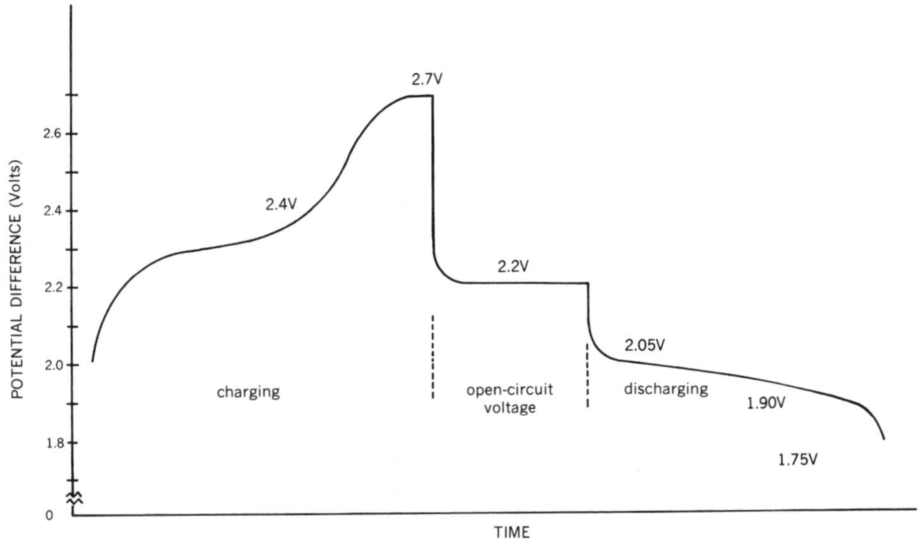

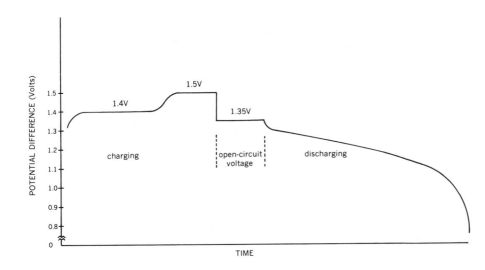

Figure 18—1
Voltage characteristics of lead-acid storage battery.

Figure 18—2
Voltage characteristics of nickel-cadmium button battery.

81

EXPERIMENT 19

Thermal Conductance of
an Insulating Material

Objective   To determine the thermal conductance of various materials.

Introduction   The finite quantity of fossil fuel laid down in the earth millions of years ago makes the conservation of energy imperative. Our grandchildren, if not we ourselves, may live to see the day when all oil and natural gas are gone and coal is scarce. One way to assure wise use of energy is to reduce the tremendous waste that occurs through oversight, carelessness, and ignorance. Obvious sources of wasted energy should be identified and the conditions corrected. It is vital for all of us to be aware of our responsibility to manage our own fossil fuel consumption frugally and with concern for others.

In a region where the heating load is over 8,000 degree-days*, like in the northern part of the United States, the heat loss from a house to the outside air should be as little as possible. We can reduce heat loss or gain through walls and roof by building in insulating materials to retard the rate of heat conduction. It is well to keep in mind that the rate of heat loss or gain can be cut in half by doubling the thickness of insulation. Adequate insulation of roof and walls, use of double windows and doors, and reduction of air leakage around the windows and doors are part of good citizenship as well as good economics. Floors over unheated spaces should be insulated. Pulling heavy drapes across the windows at night can also save a significant amount of energy. In a typically well-insulated, one-story house with one air change per hour, 27% of the heat losses were through the windows and doors, 40% through air exchange with colder outside air, 15% through the walls, 13% through the roof, and 5% through the floor. As long as a temperature difference exists between inside and outside air, heat will flow from the region of higher temperature to a region of lower temperature. It makes good sense to reduce this flow to a minimum.

The rate (quantity per unit time) at which most physical processes take place is proportional to a driving force, the nature of which depends on the kind of process. In the flow of heat by conduction, the driving force is a difference in temperature, which always moves heat from material or space at a higher temperature to material or space at a lower temperature.

---

* Degree-days are determined by subtracting the daily average outdoor temperature from 65° F and adding these differences for each day of the heating season.

Opposing the driving force is always a resistance to which the rate is inversely proportional. Therefore:

$$\text{Rate} \propto \frac{\text{driving force}}{\text{resistance}}.$$

Fourier's law states that the rate of heat flow through a material is proportional to the temperature difference and to the area, and inversely proportional to the thickness of the material. If Q is the amount of heat flowing in time t, then $\frac{Q}{t}$ is the rate of heat flow or the amount of heat flowing per unit time, and Fourier's law can be written in algebraic form as follows:

$$\frac{Q}{t} = \frac{k \cdot A \cdot \Delta T}{L},$$

where k is a proportionality constant.

Since the driving force is ΔT, the resistance R must be

$$R = \frac{L}{kA},$$

and we can visualize that R should be directly proportional to the length of heat travel and inversely proportional to the area. The proportionality constant k, known as thermal conductivity, represents a physical characteristic or property of the material through which the heat is conducted, and is expressed as

$$k = \frac{Q}{t \cdot A \cdot \Delta T / L} = \mathrm{BTU}/\mathrm{hr} \cdot \mathrm{ft}^2 \cdot \mathrm{F}°/\mathrm{ft}, \text{ or } \mathrm{BTU\text{-}ft}/\mathrm{hr} \cdot \mathrm{ft}^2 \cdot \mathrm{F}°.$$

The coefficient k is based on a material thickness of one foot parallel to the direction of heat flow, and an area of one square foot perpendicular to the direction of heat flow.

Frequently we are considering a thickness other than one foot, and the term C, the thermal conductance, is used for the particular thickness under consideration. Or,

$$C = \frac{Q}{t \cdot A \cdot \Delta T} = \mathrm{BTU}/\mathrm{hr} \cdot \mathrm{ft}^2 \cdot \mathrm{F}°.$$

When considering a wall or roof, we speak of the overall coefficient of heat transmission U for the combination of materials making up the wall or roof from one side to the other (air-wall-air). Even the thin film of air in contact with the wall offers resistance to heat transmittance and has its effect on the value of U. Or,

$$U = \frac{Q}{t \cdot A \cdot \Delta T} = \text{BTU}/\text{hr} \cdot \text{ft}^2 \cdot \text{F}°.$$

Since k is based on 1-foot thickness and 1-square-foot area in the definition equation, R becomes the reciprocal of k, or $R = \frac{1}{k}$. Likewise, R is the reciprocal of C and of U because the thickness is taken into account— $R = \frac{1}{C}$ and $R = \frac{1}{U}$. With U, it is the total resistance of the wall as it stands plus the thin film of air on each side.

In this experiment you will be able to measure for yourself the heat transmission that occurs through various materials. The experiment employs a standard engineering method, and the thermal conductance C that is determined can be used to compare the heat-conducting qualities of materials. It is the reciprocal R of the C value $\left(R = \frac{1}{C} \right)$ that measures the resistance of materials to heat transmission. This R value is commonly printed on a package of insulating material to assist the purchaser in selecting the best insulation for the money.

You can also try measuring the U factor (overall coefficient of heat transmission) for two or more materials used in combination as usually occurs in home construction—whether they be lath, plaster, and siding; insulation, sheathing, and siding; etc. You might also try the determination with an air space between two pieces of the same material. In general, when materials are used in combination, their R values are added together to find their combined insulation factor. The reciprocal of the sum is the overall coefficient of heat transmission U of the wall. You can test the validity of this yourself with this apparatus, provided adjustment is made to accommodate the thickness of the wall to be tested.

The engineering design is presented in figure 19–1 as a standard. It is suggested that you scale down this design to a more manageable size and cost. Perhaps the outside box could be 2-×-2-×-2 feet and other insulation materials could be substituted. If the C of your material is known, you can compare that value with your own C measurement and then compute a correction factor. This can then be applied to your other determinations.

The whole idea of this method is to measure the amount of heat that escapes through one wall of a container. Since the rate of heat loss depends on the temperature difference between the inside and outside of this wall, the other five sides or walls of the container must have an outside temperature equal to the inside temperature. Therefore, a box within a box is used. You can also sense that the rate of heat loss through the test wall

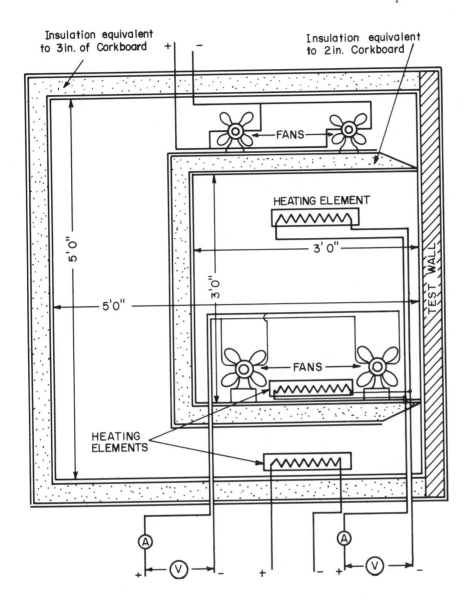

Insulation equivalent to 3 in. of Corkboard

Insulation equivalent to 2 in. Corkboard

FANS

HEATING ELEMENT

5'0"

3'0"

3'0"

5'0"

TEST WALL

FANS

HEATING ELEMENTS

A

V

V

A

Figure 19—1
Engineering test box for measuring thermal conductivity of various materials. Source: Reprinted by permission from ASHRAE *Handbook of Fundamentals,* 1972.

material depends upon its area. (Your value of C is for the "as is" thickness of the material being tested.) These factors are related as follows:

$$\frac{Q}{t} = C \cdot A \ (T_i - T_o),$$

where $\frac{Q}{t}$ =rate of heat transmittance (BTU/hour), C=thermal conductance (BTU/hour·ft²·F°), $A=$ area of material being tested in square feet (perpendicular to direction of heat flow), T_i=air temperature (°F) inside the box, and T_o=air temperature (°F) outside the box.

The amount of heat that escapes through the test wall is determined by the amount of electrical energy put into the inner box that just keeps the temperature of the box constant. This energy is equal to the electric power P times the time t that the energy is being added. The power in watts is the product of the voltage and the current applied to the heater in the inner box.

Materials

Insulating material and lumber for construction of an outer 2-×-2-×-2-foot, five-sided box. Insulating material and lumber for an inside box, two electric heaters, small electric fan, three thermometers, two ammeters, two voltmeters, two variable AC voltage transformers or two variable low voltage DC power supplies, two different insulating materials for testing.

Procedure

1. Construct the two five-sided boxes; the open sides of the two boxes are placed in the same plane with uniform spacings between all of the other corresponding walls of the inner and outer boxes.

2. Mount the small fan in the outer box so that it can gently circulate the air to maintain a uniform temperature.

3. Mount the electric heaters—one in each box. Nonluminous heaters should be used, so that uniform temperatures can be assured. If you do use a radiant heater coil, incorporate a voltage control (either a Variac for AC or a low voltage variable power supply for DC) so that you will not have to have it turned all the way on or off.

4. Make provisions for temperature measurement in the outer and inner boxes. See that any holes are sealed to prevent unwanted heat losses.

5. Make provisions for electrical connections to the heaters through the box walls and to appropriate meters. If you use AC, be sure to use AC meters.

6. Test the unit for correct operation of all components. Start the heaters one at a time at very low applied voltages and increase gradually to make sure the meters deflect correctly and are in the proper range.

7. Measure the length and width of the opening of the inner box.

8. Apply the test wall material to the open side of the apparatus. Make sure the fit is tight and no cracks are present. Perhaps some gasket material will help.

9. Bring the temperatures of the inner and outer boxes to about 10° above the outside temperature. Keep checking the readings of the inner and outer box thermometers and gradually adjust the voltage to the heater in one of the boxes until the two temperatures are the same and remain constant.

10. When this equilibrium is achieved, the rate at which electric energy is applied to the heater in the inner box is equal to the rate at which this heat energy is escaping through the test wall. Record the readings of the voltmeter V and ammeter I for the inner box. Also record the outside air temperature and the temperature of the inner box.

From these values and the area (square feet) of the inner box opening, you can compute the C value of the material from

$$C = \frac{V \times I \times 3.413}{A\ (T_i - T_o)},$$

since

$V \times I \times 3.413$ (rate of energy input) $= \dfrac{Q}{t}$ (rate of energy loss in BTU/hr).

It is important to keep track of the units involved here. The constant 3.413 is the conversion factor required to convert the watts (volts × amperes) you measured to BTUs per hour.

11. Repeat this procedure several times, increasing the applied voltage to the heaters so that the inner box temperature will be about 20°, 30°, 40°, etc., above room temperature. These ΔT's do not have to be exactly in 10° increments. Accept and measure whatever values you have when thermal equilibrium has been achieved.

12. Plot a graph of your computed C value of the material versus the ΔT that was used.

Results

1. What is the shape of the trend line of your graph? What does it show?

2. Examine the variation of your plotted points from the trend line. This will give you a good idea of how consistent your experimental technique is.

3. If you compare the reciprocal of the C value you measured with the R value for your test material, you can come up with a calibration for your test apparatus.

4. Test another material and apply your correction factor. Now how do the R values compare?

5. Try various materials in combination. Check the construction of a wall in your own home and determine its R value. This could be done by referring to tables of C or R values in reference 3, or even experimentally.

References

1. Farrington Daniels, *Direct Use of the Sun's Energy*, Yale University Press, 1964.

2. Warren L. McCabe and Julian C. Smith, *Unit Operations of Chemical Engineering*, McGraw-Hill, 1956.

3. *ASHRAE Handbook of Fundamentals*, 1972.

4. *Standard Code for Heat Transmission through Walls*, ASHVE Trans., Vol. 34, 1928.

CLASSROOM ACTIVITY A

Energy Consumption

Introduction   Solar energy is used for heating and cooking in remote areas of the world and to a limited extent for home heating in more developed countries. A small solar-powered generator made it possible for men to survive and communicate by radio and television from the moon.

The sun's energy is partially absorbed by the atmosphere (and less and less gets through the more we pollute the air), but from the small portion that does get through, it has been estimated that each square mile of the earth's surface could produce 180,000 kilowatts of power, at only a 10% conversion efficiency. For example, if solar-powered generating plants covered the 3.2 million acres laid waste by strip mining, we could be producing about 1 billion kilowatts of power from them. This is about 100 times the power New York City is expected to need in the year 2000. Another possibility would be satellite solar power stations transmitting power to earthbound receivers by microwave. Such stations being above our atmospheric envelope, losses from atmospheric absorption of solar radiation and cloud cover would be avoided.

While these examples of power generation may seem fanciful, they are within the realm of future feasibility—the technology is in hand now. We can perhaps broaden interest in developing greater use of solar energy by demonstrating on a very small scale that water can be heated without concomitant production of pollution. On the other hand, if we burn oil or gas for the same purpose, we produce polluting substances at home; if we use electricity, the local power plant is required to burn more fuel, creating extra pollution there.

Many might wonder how water can be heated by solar energy at night and during stormy weather. It can't be. But, by storing solar-heated water in a 100-gallon insulated tank, a large part of the problem can be solved. If this doesn't supply the family's demand for hot water, the conventional oil, gas, or electric water heater goes to work until the sun can again take over its job.

Our nation's energy production has had a difficult time keeping up with the growing demand for energy. So much of this energy is from fossil fuels that great concern exists regarding exhaustion of these finite resources. Man has inhabited spaceship earth for thousands of years, yet most of the fuels and minerals mined or pumped from the earth have been removed and consumed within the last century. Unless conservation of fuel and mineral resources starts soon, our descendants may have rough sledding. To get the most from the alternate source of the sun requires strict attention to good conservation practices, which will also pay off when other sources of energy are used. This involves such things as extra insulating of buildings, doubling up with car pools, using less than a full tub of water when taking a bath and keeping showers under 20 minutes, turning off unused lights and appliances, using the clothesline instead of the dryer, riding bikes for short trips rather than driving, turning thermostats down—the opportunities for energy conservation go on and on. A national consciousness of the need for conservation must come about before significant reduction in fossil fuel consumption can be realized.

Objective

This first activity will give you a realistic picture of your own energy consumption. When you read about solutions to energy problems, you will be able to relate the amount of energy you use to the nation's patterns of consumption. As you grow older, you hopefully will run your family activities and make decisions at your job with conservation principles in mind, both for reasons of good stewardship on our planet and for economic reasons.

Procedure

Keep track of your own energy consumption day by day. By actually measuring it and recording your data, you will be able to observe daily and seasonal variations.

For example:

1. Keep a daily record of your family's electric meter readings. Keep in mind that for every kilowatt-hour of electrical energy you use, the power company consumes 3 kilowatt-hours of fuel energy.

2. Keep a record of the amount of fuel oil required to heat your home. Relate this to the heating degree-day variations throughout the season. Weather broadcasts frequently give this information. See "Determining Heating Degree-Days per Year," experiment 3.

3. Make daily readings of your gas meter, and relate these to specific uses. When does the maximum consumption occur?

4. How much energy in the form of gasoline does your family use? You can determine this from the gas mileage you get from your car and from the fact that gasoline gives about 125,000 BTUs of energy per gallon.

5. Determine the amount of heat you actually use when taking a shower or tub bath. Which requires more water? Compare the amount of energy required for a full tub and for a tub with no more water than necessary. More detailed instructions for this are included in "Energy Consumed by a Hot Bath," experiment 13.

Table A–1   **AVERAGE HEATING VALUES OF FOSSIL FUELS**

Natural Gas .... 1,000 BTU/ft³

Oil .... 140,000 BTU/gal (home heating)
150,000 BTU/gal (industrial)

Anthracite .... 12,500 BTU/lb

6. Keep a record of the cost for each of your own energy-using activities. Assume the following rates: electricity, 3¢/kwh; gasoline, 60¢/gal; fuel oil, 40¢/gal; natural gas, $1.90/1,000 ft³.

7. See if you can figure out how much energy is required to cook various meals. You can use the technical data (voltage and current ratings) that come with your range or BTU rating of your gas stove. These ratings are frequently marked on the serial number plate.

Energy Demand

The basis for the energy problem is our reliance on a finite amount of fossil fuel. As we awaken to our environmental responsibilities and obligations to future generations, the conservation of all of our resources must become a way of life for everyone. By analyzing data about the past, such as Tables B–1, B–2, and B–3, we can determine which changes need to be made to avoid major catastrophies in the future. These tables include a summary of

Table B–1

TOTAL U.S. ENERGY AND ELECTRICITY CONSUMPTION, GROSS NATIONAL PRODUCT, AND POPULATION— SELECTED YEARS, 1920–1970

Year	Total energy consumption (trillion BTU)	Electricity consumption (billion kwh)	GNP (billion 1958 dollars)	Population (million)	Per capita			Per $1 of GNP	
					Energy consumption (million BTU)	Electricity consumption (kwh)	GNP (1958 dollars)	Energy consumption (thousand BTU)	Electricity consumption (kwh)
1920	19,782	57.5	140.0	106.5	185.8	540	1,315	141.3	0.41
1930	22,288	116.2	183.5	123.1	181.1	944	1,490	121.5	0.63
1940	23,908	182.0	227.2	132.6	180.3	1,376	1,720	105.2	0.80
1950	34,154	390.5	355.3	152.3	224.3	2,564	2,342	96.1	1.10
1960	44,960	848.7	487.7	180.7	248.8	4,967	2,699	92.2	1.74
1965	53,785	1,157.4	617.8	194.6	276.4	5,948	3,175	87.1	1.87
1970	68,810	1,648.3	724.1	205.4	335.0	8,025	3,525	95.0	2.28

Notes and Sources: Total energy consumption, 1920–1940, from U.S. Bureau of the Census, *Historical Statistics of the United States, Colonial Times to 1957* (Washington: GPO, 1960); 1950–1960, from U.S. Bureau of Mines Information Circular 8384, *An Energy Model for the United States* (Washington: GPO, 1968); 1965 from U.S. Bureau of Mines, *Minerals Yearbook, 1969* (Washington: GPO, 1970); 1970 from U.S. Bureau of Mines News Release dated March 9, 1971.

The data on electricity consumption represent net generation by privately and publicly owned utilities as well as other generation (e.g., industrial firms' own electricity production) and, in addition, include net imports of power. The 1920–1965 data are from the U.S. Bureau of the Census and the Federal Power Commission, shown in Edison Electric Institute, *Statistical Yearbook of the Electric Utility Industry for 1969* (New York, 1970) and *Historical Statistics of the Electric Utility Industry* (New York, 1963). Figure for 1970 is preliminary estimate.

GNP for 1920 from U.S. Bureau of the Census, *Long-Term Economic Growth 1860–1965* (Washington: GPO, 1966); 1930–1965, U.S. Department of Commerce data shown in *Economic Report of the President, January 1971* (Washington: GPO, 1971); 1970 figure from U.S. Department of Commerce, *Survey of Current Business,* May 1971.

Population for 1920–1965 from U.S. Bureau of the Census, *Statistical Abstract of the United States* (Washington: GPO, annual); for 1970 from *Economic Report of the President, January 1971.*

total U.S. energy and electricity consumption and other data for every tenth year since 1920.

By analyzing these data it may be possible to suggest what will happen in the near and even distant future if the indicated trends continue. Such data are used to predict future demands and are the basis for management decisions by those responsible for anticipating and meeting the future demands.

1. Plot a graph that shows the trend of U.S. energy consumption since 1920.

2. Plot any of the other data on the same or separate axis, in order to help you visualize the extent of the energy problem.

Table B–2

TOTAL U.S. ENERGY CONSUMPTION, BY SOURCE AND FORM OF USE—SELECTED YEARS, 1920–1970

Year	By source [a]					By form used			
						Fuel and power			As raw material
	Coal	Natural gas	Petroleum	Hydro and nuclear	Total	Total	Electricity	Other	
					Trillion BTU				
1920	15,504	827	2,676	775	19,782	. . . . .	1,663	. . . . .	. . . . .
1930	13,639	1,969	5,898	785	22,288	. . . . .	1,965	. . . . .	. . . . .
1940	12,535	2,726	7,781	917	23,908	. . . . .	2,458	. . . . .	. . . . .
1950	12,914	6,150	13,489	1,601	34,154	32,712	5,142	27,570	1,442
1960	10,414	12,699	20,067	1,780	44,960	42,715	8,387	34,328	2,245
1965	12,358	16,098	23,241	2,088	53,785	51,140	11,104	40,036	2,645
1970	13,792	22,546	29,617	2,855	68,810	64,910	16,967	47,943	3,900
					Percent				
1920	78.4%	4.2%	13.5%	3.9%	100.0%	. . . .	8.4%	. . . .	. . . .
1930	61.2	8.8	26.5	3.5	100.0	. . . .	8.8	. . . .	. . . .
1940	52.4	11.4	32.4	3.8	100.0	. . . .	10.3	. . . .	. . . .
1950	37.8	18.0	39.5	4.7	100.0	95.8%	15.1	80.7%	4.2%
1960	23.2	28.2	44.6	4.0	100.0	95.0	18.7	76.3	5.0
1965	23.0	29.9	43.2	3.9	100.0	95.1	20.6	74.4	4.9
1970	20.0	32.8	43.0	4.1	100.0	94.3	24.7	69.7	5.7

Sources: Same as the sources shown for total energy consumption in Table B–1 except that data on energy consumed in electric generation were based, for 1920, on data in U.S. Bureau of the Census, *Historical Statistics of the United States*, op. cit.; and for 1930–1940 was obtained on the basis of data shown in Edison Electric Institute, *Historical Statistics of the Electric Utility Industry* (New York, 1963). The EEI series on total net electric generation (expressed in kwh) was multiplied by the EEI estimate of the heat rate (BTU per kwh) for all boiler fuels; the result was added to the hydro figures, shown above.

[a] Coal includes bituminous coal, anthracite, lignite; petroleum includes natural gas liquids; the nuclear component (not shown separately) amounted (in trillion BTU) to 6 in 1960, 39 in 1965, and 208 (or 0.3% of total energy consumption) in 1970.

Put yourself in the shoes of the president of an electric power company whose plant is producing at about 90% capacity. Or perhaps you might be an advisor to the President of the United States on energy matters, or just consider yourself an ordinary resident of this planet. What energy-related questions should you be concerned with in order to maintain the future well-being of your company, your country, yourself?

What other data do you need in order to get a clearer picture of what is required of us in the future?

Table B–3

CONTRIBUTIONS OF DIFFERENT COMPONENTS TO CHANGES IN U.S. ENERGY CONSUMPTION, 1960–1965 AND 1965–1970

End-use sector and type of energy consumed	1960 share of total U.S. energy consumption (percent)	1960–1965		1965 share of total U.S. energy consumption (percent)	1965–1970		Change between 1960–1965 and 1965–1970 in the contribution of different component (col. 6 minus col. 3)
		Average annual percent rate of change	Contribution to average annual percent rate of change in total energy consumption [a]		Average annual percent rate of change	Contribution to average annual percent rate of change in total energy consumption [a]	
	(1)	(2)	(3)	(4)	(5)	(6)	(7)
Household and commercial:							
Fossil fuels	22.7	3.05	0.69	22.1	3.51	0.78	0.09
Electricity	2.8	9.08	0.25	3.6	8.31	0.30	0.05
Industry:							
Fossil fuels	33.1	3.37	1.12	32.6	3.78	1.23	0.11
Electricity	2.9	4.58	0.13	3.0	6.84	0.21	0.08
Transportation:							
Fossil fuels	24.1	3.28	0.79	23.6	5.27	1.24	0.45
Electricity conversion losses	12.9	5.29	0.68	14.0	9.42	1.32	0.64
Total energy	100.0	3.65	3.65	100.0	5.05	5.05	1.40
In fuel and power uses	95.0	3.67	3.49	94.9	4.88	4.63	1.14
In nonenergy uses	5.0	3.33	0.17	5.1	8.07	0.41	0.24

Source: U.S. Bureau of Mines, *Minerals Yearbook* and various U.S. Bureau of Mines releases.

Note: Components of cols. (1), (3), (4), (6), and (7) do not quite add to totals because of rounding and because of omission of electricity consumption in Transportation (a statistically insignificant sector) and a minor "miscellaneous and unallocable" component.

[a] Weighted by relative contribution to nationwide energy consumption in 1960 and 1965, respectively, as shown in cols. (1) and (4).

CLASSROOM ACTIVITY C

Solar Constant

Introduction The sun is quite uniform in the rate it gives off energy, varying perhaps less than 5% over long periods of time. This energy comes from nuclear fusion reactions that take place within it. While these reactions are very complex, in summary it may be said that four hydrogen nuclei combine to form one helium nucleus. The earth intercepts some of this energy. The solar constant is defined as the mean value over the whole year of the intensity of solar radiation on a surface placed normal (perpendicular) to the sun's rays just outside the earth's atmosphere. It is usually accepted as being 1.94 cal/cm²·min (7.16 BTU/ft²·min), but there is some doubt as to its exact value. It may be as much as 2.0 cal/cm²·min, or 2.0 langleys/min.

Objective As the earth moves around the sun, its distance varies between 91.5×10^6 miles and 94.5×10^6 miles (147×10^6/km and 152×10^6 km). We are closest to the sun on January 4th and farthest away on July 4th. As a result, the amount of solar energy received by the earth's atmosphere is not constant, but is actually somewhat larger in winter in the Northern Hemisphere. The purpose of this activity is to show you the extent of this variation, and to see to what extent it will affect our attempts to utilize the sun's energy.

Data Table C–1 shows the date, the solar radiation intercepted by the earth outside the atmosphere on that date, and the correction of that value to obtain the mean solar constant of 1.94 cal/cm²·min or 428.7 BTU/ft²·hr. Plot the values of the solar constant versus the date of the month for every date given. Care will have to be taken in choosing your scale so that the variations in the solar constant values can be seen. If you wish, the variations from the mean values could also be plotted.

Discussion questions 1. What is the shape of the plotted graph? Why is it shaped this way?

2. What does the graph show?

3. What is the percent variation in the solar constant from its maximum to minimum values?

4. At what times of the year does the solar constant change at the greatest rate? Why?

5. How will this variation affect the design of solar energy devices?

94

6. What factors do you suppose will have a more significant effect on the practical use of solar energy?

Table C–1

SOLAR CONSTANT AND REDUCTION OF MEASURED SOLAR RADIATION INTENSITY TO AVERAGE SOLAR DISTANCE

Date		Solar Constant (cal/cm² min)	Red. to Mean Distance (%)	Date		Solar Constant (cal/cm² min)	Red. to Mean Distance (%)	Date		Solar Constant (cal/cm² min)	Red. to Mean Distance (%)
Jan.	1	2.007	−3.32	May	1	1.910	+1.55	Sept.	1	1.905	+1.83
	6	2.007	−3.31		6	1.906	+1.80		6	1.910	+1.58
	11	2.006	−3.28		11	1.901	+2.04		11	1.915	+1.32
	16	2.005	−3.23		16	1.897	+2.26		16	1.920	+1.05
	21	2.003	−3.16		21	1.894	+2.46		21	1.925	+0.77
	26	2.001	−3.06		26	1.890	+2.65		26	1.930	+0.40
	31	1.999	−2.93		31	1.887	+2.82	Oct.	1	1.936	+0.21
Feb.	1	1.999	−2.90	June	1	1.886	+2.85		6	1.942	−0.08
	6	1.995	−2.75		6	1.884	+2.99		11	1.947	−0.37
	11	1.991	−2.58		11	1.882	+3.11		16	1.953	−0.66
	16	1.987	−2.39		16	1.880	+3.22		21	1.958	−0.94
	21	1.983	−2.18		21	1.878	+3.29		26	1.963	−1.21
	26	1.979	−1.95		26	1.877	+3.31		31	1.969	−1.47
Mar.	1	1.976	−1.80	July	1	1.877	+3.37	Nov.	1	1.970	−1.52
	6	1.971	−1.56		6	1.876	+3.38		6	1.975	−1.77
	11	1.966	−1.30		11	1.877	+3.36		11	1.980	−2.01
	16	1.960	−1.02		16	1.878	+3.31		16	1.984	−2.23
	21	1.954	−0.74		21	1.879	+3.24		21	1.988	−2.44
	26	1.949	−0.46		26	1.881	+3.15		26	1.992	−2.62
	31	1.944	−0.17		31	1.883	+3.03	Dec.	1	1.996	−2.79
April	1	1.943	−0.11	Aug.	1	1.883	+3.01		6	1.998	−2.94
	6	1.937	+0.17		6	1.886	+2.87		11	2.001	−3.07
	11	1.931	+0.46		11	1.889	+2.71		16	2.003	−3.17
	16	1.926	+0.74		16	1.892	+2.53		21	2.005	−3.25
	21	1.920	+1.02		21	1.896	+2.33		26	2.006	−3.30
	26	1.915	+1.29		26	1.900	+2.11		31	2.007	−3.31

Source: Pollack, 1958: private communication.

Notes: Average solar constant: 1.94 cal/cm²·min.
The deviations of the various years are less than 0.1%.

CLASSROOM ACTIVITY D

Insolation

Introduction When solar radiation strikes a surface or boundary between two regions, three things happen to the energy: absorption, reflection, and transmission. Sometimes one process is dominant over the others as in the case of mirrors, transparent glass, or opaque bodies. Yet, technically, all three do occur at

Table D–1

HOURLY RADIATION FOR THE MONTH OF AUGUST 1972

Date	5	6	7	8	9	10	11	12	13	14	15	16	17	18	19	20	TOTAL
1		2	16	34	50	59	65	69	67	60	50	42	25	11	2		550
2		1	13	27	41	45	46	62	57	60	50	34	14	7	2		439
3		1	0	14	42	38	43	48	33	32	18	22	14	2	1		317
4				12	40	33	66	63	63	64	57	24	6	6	5		440
5		3	17	37	51	60	64	66	68	72	62	45	29	15	3		593
6		4	19	34	49	61	69	70	70	66	46	30	31	13	3		573
7		1	3	2	4	19	36	23	24	19	7	5	15	8			165
8		1	11	32	46	56	68	72	69	55	45	22	14	11	2		504
9		1	10	22	19	34	34	30	44	40	44	32	8	8	2		330
10		4	20	36	50	63	61	39	55	57	49	45	16	13	3		509
11		3	18	28	34	54	68	68	72	57	39	36	24	13	2		516
12			6	11	10	14	16	22	18	16	34	24	16	8	1		195
13			11	15	20	20	34	54	62	61	52	37	21	11	2		402
14		2	15	28	41	54	58	46	52	50	48	31	16	5	1		447
15			9	7	20	15	25	17	38	46	50	39	17	6	2		291
16		2	17	33	46	59	61	34	42	47	42	29	17	7	1		438
17			1	3	7	9	13	9	14	14	9	9	5	2			93
18			3	16	4	27	48	60	56	51	31	28	19	7	1		348
19		1	17	36	50	61	74	73	56	62	45	34	21	8	1		539
20		1	15	31	45	56	66	74	71	67	58	44	27	14	3		571
21		1	14	29	42	52	61	65	64	60	52	39	24	10	1		514
22		1	10	24	41	52	58	65	61	56	47	35	23	9	1		482
23			8	26	39	48	55	59	56	49	41	30	17	5			434
24			5	18	33	43	51	56	54	45	36	13	12	5			372
25			6	18	30	40	47	51	20	6	31	32	16	2			297
26			5	16	27	29	29	46	52	43	20	6					273
27			3	12	39	47	54	43	47	27	38	23	17	4			353
28		1	3	19	43	57	35	33	47	48	30	28	18	7	2		371
29		1	8	17	38	52	63	66	64	59	47	35	21	8	2		478
30		1	9	25	40	53	60	65	62	55	46	33	21	7	2		477
31		1	9	25	38	51	56	61	62	56	46	33	20	8	1		465
No. of days		21	30	31	31	31	31	31	31	31	31	31	30	30	24		31
Mean		2	10	22	35	44	51	51	52	48	41	30	18	8	2		412

Note: Location: New York City.

Unit: Langleys on a horizontal surface.

each boundary. An important boundary or interface as far as we are concerned is the atmosphere. The solar constant activity shows the small variability that occurs without any atmospheric effects. Activity D will examine some of the effects that the atmosphere has on the radiation that reaches the earth's surface.

During transmittance through the atmospheric layers, the sun's radiation is scattered and absorbed by dust, gas molecules, ozone, and water vapor; the extent of this depletion depends upon the composition of the atmosphere and the distance the rays have to travel through it to reach the earth's surface. A summary of what happens to this incoming radiation is seen in figure D–1.

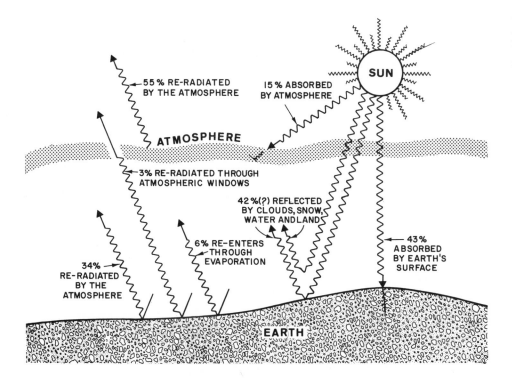

Figure D—1
Illustration showing what happens to incoming solar radiation.

Figure D–2 shows another way to indicate what is happening to the solar radiation as it interacts with our atmosphere. The dashed line indicates what might be expected if the atmosphere did not absorb the energy. The intensity of the energy reaching the earth would follow a smooth curve when related to the wavelength of radiation. Only a small part of this graph covers the visible part of the spectrum—from about 4,000 angstroms (4×10^{-7}m) on the blue side to about 7,000 angstroms (7×10^{-7}m) on the red side. You can see from this graph that much of the solar energy lies in the infrared region (wavelengths longer than we can see with our eyes), while at the same time the most intense radiation received does just include that to which our eyes are sensitive (the red), around 7,000 angstroms. While we cannot see the infrared radiation, we can certainly feel it with our skin and get the sensation of heat. The higher energy ultraviolet and x-radiation (wavelengths less than 4,000 angstroms) tend to be absorbed by the oxygen, nitrogen, and ozone molecules in the outer atmosphere (fortunately so for living things). It is these rays that cause sunburn and other harmful effects. The infrared radiation tends to be absorbed at specific wavelengths by carbon dioxide, water vapor, water droplets, and ozone. Thus, there is not the smooth curve, but an irregular curve showing the decrease in intensities reaching the earth, which results from atmospheric absorptions.

Figure D—2

Spectral distribution of direct solar radiation at normal incidence for the upper limit of the atmosphere and at the earth's surface for conditions at sea level, 30 mm of precipitable water, and 400 dust particles per cc.

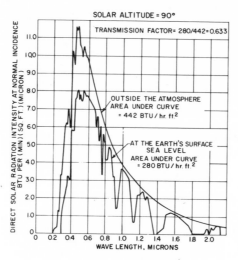 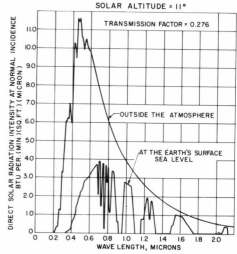

Data

This activity asks you to plot the values of solar energy received on a horizontal surface versus the hour of day at New York City during August 1972 for several different days (see table D–1). The values are in langleys/hr—that is, the amount of energy (in calories) received per unit area (in square centimeters) during each hour. Can you tell why the values differ? Plot the mean values for the month for each hour of day on the same axis.

Discussion questions

1. How does the plot of the mean values differ from the plot of the daily values? What causes the differences?

2. How do the hourly values change during the day? Why?

3. How would you expect the January values to differ from the August values? Why?

4. How would the plots differ if the receiving surface were always perpendicular to the sun's rays, rather than horizontal?

5. How does the thickness of the atmosphere through which the rays travel vary in the morning and evening as compared to noon? What effect does this have on the amount of energy reaching the surface? Can you see this on your graph?

6. Are the effects of cloud cover evident in your graph? How do they appear on the graph? Why?

7. Why do the incoming values of solar radiation at the surface at noon on clear days at middle latitudes vary around 1 langley per minute, while the solar constant values are measured to be about 2 langleys per minute?

Solar Energy Collection

Table E–1 shows the quantity of solar energy received by different areas, assuming the average intensity of radiation on a horizontal surface is 1 cal/cm²·min. Try the following problems. They should help you get a feel for the amount of energy available by using the sun.

a. Why is this value of average intensity different from the solar constant?

b. How much energy would a small-model solar collector 1½-×-2½ feet receive in one day, assuming 500 minutes per day of solar radiation?

1. Assuming the collector is 30% efficient, what is the amount of usable heat produced?

2. Assuming an average electrically heated household uses 30 kwh of energy per day for heating, what percent of this could be supplied by the solar collector?

3. How much fuel is being saved by using this collector? Assume an oil-fired plant produced your electricity, that residual fuel oil produces 150,000 BTU/gallon, and that 1 BTU=252 calories. This also assumes, of course, that the power plant can reduce its input directly in proportion to the reduced output or the reduced consumption of electric energy at point of use. You will probably find there would have to be thousands of such savings before the power plant could actually shut down a boiler. Remember that for every unit of electric energy consumed in the home, three times as much fuel energy is consumed at the power plant to develop and transmit the energy to your home, since about 70% of the fuel energy is lost as heat at the plant.

4. Now let's use some more realistic figures. The average insolation on a sunny day for Albany, New York, on a horizontal surface in January is 542 BTU/ft² per day (or 1,165 BTU/ft² per day on a 45°-sloped collector). If the collector is 35% efficient, this would give a usable 410 BTUs per day for each square foot of collector area or 112 langleys per day. From this value, compute the amount of energy you would have available each day in January to heat your home with a 15-×-25-foot collector. If an average house requires 102,400 BTUs per day for heating, is this size of collector sufficient to furnish all of the heat requirements on a sunny day?

Table E-1

QUANTITY OF SOLAR RADIATION RECEIVED BY DIFFERENT AREAS

(Average Intensity of Radiation is: 1 cal·cm^{-2}·min^{-1})

Area	langleys	kcal min^{-1}	kcal day^{-1}*	BTU hr^{-1}	kw (heat)	hp (heat)
1 cm^2	1.0	0.001	0.500	0.238	7.00×10^{-5}	9.39×10^{-5}
1 ft^2	929	0.929	464	221	0.065	0.087
1 m^2	10^4	10	5.0×10^3	2380	0.700	0.938
100 m^2 (roof)	10^6	10^3	5.0×10^5	2.38×10^5	70.0	93.8
1 acre	4.05×10^7	4.05×10^4	2.02×10^7	9.64×10^6	2.83×10^3	3.79×10^3
1 km^2	10^{10}	10^7	5×10^9	2.38×10^9	7.00×10^5	9.38×10^5
1 mile^2	2.59×10^{10}	2.59×10^7	1.3×10^{10}	6.15×10^9	1.81×10^6	2.42×10^6

Source: Farrington Daniels, *Direct Use of the Sun's Energy,* Yale University Press, New Haven, 1964.

Note: Conversion factors: 1 kcal=1.000 cal; 1 BTU=0.252 kcal, 1 kw=14.3 kcal·min^{-1}; 1 hp=0.742 kw; 1 ft^2=929 cm^2; 1 acre=43,560 ft^2. A complete list of conversion factors is given in the Appendix.

* Assuming 500 min·day^{-1} of solar radiation.

Solar Radiation

Should you keep a record of the daily solar radiation in your location, you would see that the variations that occur would be quite great; radiation intensity depends on many factors, such as the time of day, the season of year, the atmospheric conditions including cloud cover, humidity, etc. In spite of these variations, average values could be determined from such measurements for each hour of the day. Monthly averages could also be determined for each hour. When the results of these determinations are plotted on a graph, a family of curves results as shown in figure F–1. The

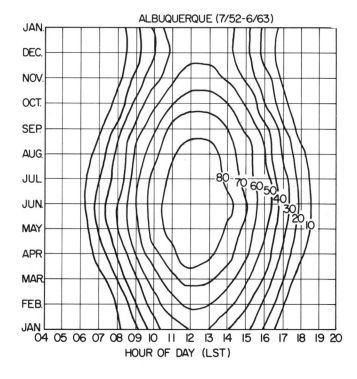

Figure F—1
Mean hourly total radiation (langleys) by months. Source: *Solar Heating and Cooling of Buildings*, Phase O, Final Report TWR, No. 25168.002.

isolines that are shown give the mean hourly total radiation in langleys for each month at Albuquerque, New Mexico, for the heating seasons of 1952 to 1963. One langley is equal to a radiation intensity where one calorie of energy falls on a surface that is one square centimeter in area. This is equivalent to 3.69 BTUs per square foot. It is from such data as these that solar energy collector systems are designed.

101

Discussion questions

1. What is the main cause of the symmetry of these curves around the 12:30-P.M. axis?

2. Why is this symmetry not around the 12-noon axis?

3. How do you account for the symmetry about the June axis?

4. What would cause this latter to shift toward the middle of May, particularly in the later hours of the day? Hint: check the mean monthly precipitation for the Albuquerque area.

CLASSROOM ACTIVITY **G**

Solar Demonstration Residence

Table G–1

MONTHLY HEATING REQUIREMENTS FOR A WEST VIRGINIA RESIDENCE

Month	Degree-days	Avg. Monthly Heat Loss	Domestic Hot Water	Total Thermal Load	Solar * Contri- bution	Auxiliary Required
January	964	11,920,824	822,182	12,743,006	9,988,944	2,754,062
February	840	10,387,440	742,616	11,130,056	9,285,696	1,844,360
March	704	8,705,664	822,182	9,527,846	10,718,064	. . . .
April	357	4,414,662	795,660	5,210,324	10,143,000	. . . .
May	115	1,422,090	822,182	2,244,272	11,082,624	. . . .
June	9	111,294	795,660	906,954	9,878,400	. . . .
July	0	0	822,182	822,182	11,210,220	. . . .
August	0	0	822,182	822,182	12,395,040	. . . .
September	66	816,156	795,660	1,611,816	11,995,200	. . . .
October	298	3,685,068	822,182	4,507,250	12,559,092	. . . .
November	603	7,456,698	795,660	8,252,358	10,354,680	. . . .
December	930	11,500,380	822,182	12,322,562	9,314,508	3,008,054
TOTALS	4,886	60,420,276	9,680,530	70,100,808	128,925,468	7,606,475

Sources: Richard Rittleman, *Solar Heating and Cooling for Buildings* Workshop, Washington, D.C., 21–23 March 1973, Part I, Technical Sessions, 21–22 March; NSF/RANN–73004, July 1973.
* Assumes infinite storage characteristics.

Table G–1 shows the monthly heating requirements and additional data regarding a solar-heated demonstration residence in West Virginia. From a study of these data, you can get an idea of the order of magnitude of the domestic energy use in this part of the country, and how much of this energy might be supplied by a solar-augmenting heating system. For a comparison, the engineering data in tables G–2 and G–3 were prepared for a proposed residence near Albany, New York.

1. Plot a graph with the months on the *x*-axis, showing the annual variation in heating degree-days for each set of data.

2. Plot a graph for each set of data showing the monthly variation in the amount of heat required (thermal load). On the same axis, plot the amount of heat contributed by collecting solar energy. (This is a monthly average figure with a collection efficiency of 35%.)

    a. During what months is an auxiliary heating system required?

b. What percent of the total energy required is supplied by the sun for this residence?

3. Compare your graphs for 2., (preceding) with figure G–1, which gives the monthly energy requirements and solar energy collection for a residence in Columbus, Ohio. How do they differ? Why do they differ? During what months is there a solar energy deficit? How could you utilize the 20% summer surplus?

Table G–2

MONTHLY HEATING REQUIREMENTS AND FULFILLMENT FOR A RESIDENCE NEAR ALBANY, NEW YORK (510 ft² @ 16,000 BTU/HDD)

Month	Est. HDD	Avg. Monthly Heat Loss	Domestic Hot Water	Total Thermal Load	Mean Daily Collection at 42.5° N Lat.	Collection daily for 510 ft²	Solar Collection per mo.	Auxiliary Fuel BTU/mo.
Jan.	1,350	21,600,000	2,320,000	23,920,000	410	209,000	6,580,000	17,340,000
Feb.	1,200	19,200,000	2,100,000	21,300,000	435	222,000	6,220,000	15,080,000
Mar.	1,000	16,000,000	2,320,000	18,320,000	550	281,000	8,710,000	9,610,000
Apr.	560	9,000,000	2,250,000	11,250,000	585	298,000	8,940,000	2,310,000
May	240	3,900,000	2,320,000	6,220,000	545	278,000	8,630,000	. . . .
June	50	800,000	2,250,000	3,050,000	555	283,000	8,480,000	. . . .
July	. . . .	. . . .	2,320,000	2,320,000	630	321,000	9,950,000	. . . .
Aug.	20	320,000	2,320,000	2,640,000	685	349,000	10,820,000	. . . .
Sept.	140	2,200,000	2,250,000	4,450,000	600	306,000	9,180,000	. . . .
Oct.	440	7,000,000	2,320,000	9,320,000	525	268,000	8,300,000	1,020,000
Nov.	770	12,300,000	2,250,000	14,550,000	400	204,000	6,120,000	8,430,000
Dec.	1,200	19,200,000	2,320,000	21,520,000	380	194,000	6,010,000	15,510,000
				138,860,000=1000 gal oil			97,940,000	69,300,000= 495 gal oil

Sources: Adapted from Richard Rittleman, *Solar Heating and Cooling for Buildings* Workshop, Washington, D.C., 21–23 March 1973, Part I, Technical Sessions, 21–22 March; NSF/RANN–73004, July 1973.

Note: $100.0 \cdot \frac{69.3}{138.86} = 100 - 50 = 50\%$ from sun if all solar heat needed can be used at 100% efficiency.

Table G–3

MONTHLY HEATING REQUIREMENTS AND FULFILLMENT FOR A RESIDENCE NEAR ALBANY, NEW YORK (690 ft² @ 16,000 BTU/HDD)

Month	Total Thermal Load ×10⁶	Collection Daily 690 ft²×10⁶	Solar Collection per mo. ×10⁶	Auxiliary Fuel BTU/mo.×10⁶
Jan.	23.92	.282	8.74	15.18
Feb.	21.30	.300	8.40	12.90
Mar.	18.32	.379	11.74	6.58
Apr.	11.25	.403	12.10	....
May	6.22	.376	11.65	....
June	3.05	.383	11.49	....
July	2.32	.435	13.48	....
Aug.	2.64	.473	14.65	....
Sept.	4.45	.414	12.40	....
Oct.	9.32	.362	11.22	....
Nov.	14.55	.276	8.28	6.27
Dec.	21.52	.262	8.13	13.39
	138.86		132.28	54.320=390 gal. oil

Sources: Adapted from Richard Rittleman, *Solar Heating and Cooling for Buildings* Workshop, Washington, D.C., 21–23 March 1973, Part I, Technical Sessions, 21–22 March; NSF/RANN–73004, July 1973.

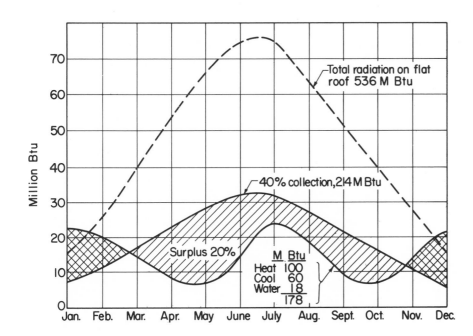

Figure G—1
Solar energy supply and requirements of a residence in Columbus, Ohio. Source: Richard Rittleman, Solar Heating and Cooling for Buildings Workshop. Washington, D.C., March 21-23, 1973, Part I, Technical Sessions, March 21-22; NSF-RANN-73004, July 1973.

Thermal Insulation

Introduction The need for energy conservation is clear, since fossil fuel resources are finite and are being rapidly depleted. Blanket-type fiberglass and mineral wool insulation are commonly used to reduce heat losses in the home, and they not only make economic sense as the cost of fuel increases, but constitute good environmental stewardship as well. This activity shows quantitatively just how much energy is saved by various thicknesses of insulation used. By obtaining the current costs of the insulation from your hardware store or building supplier, you could also determine the total cost of insulation that is appropriate for your own home.

The R value printed on the package of insulating material measures the resistance of a material to heat transmission. Thus, the greater the R value, the better the insulating effect of the material. The reciprocal of R is called thermal conductance C, and this value is the rate of heat loss in $BTU/hr \cdot ft^2 \cdot F°$ for the thickness of insulation in the package.

Table H–1 ***R* VALUES FOR VARIOUS THICKNESSES OF BLANKET-TYPE FIBERGLASS INSULATION**

Thickness (inches)	R Value
1	3.70
2	7.40
3	11.10
4	14.80
5	18.50
6	22.20
7	25.90
8	29.60
9	33.30
10	37.00
11	40.70
12	44.40

Procedure Table H–1 shows the R values for various thicknesses of blanket-type insulation. From this information compute the C value for each thickness using $C = \dfrac{1}{R}$. Plot these C values on the y-axis with the thickness of insulation in inches on the x-axis.

1. How do thermal conductance values change with thickness?

Discussion questions

2. What does the plot of R indicate?

3. Compare the amount of heat loss using 6 inches of insulation with that associated with 3 inches. Compare 12 inches with 6 inches. What do you observe?

4. Try plotting the R values and then the C values versus thickness on 2-×-2 cycle log paper. Note the difference in the shape of the curves. What does this indicate about the relationship between C and R?

Glossary of Solar Energy Terms

The ratio of the incident radiant energy absorbed by a surface to the total radiant energy falling on the surface. — ABSORPTIVITY

The ratio of the light reflected by a surface to the light falling on it. — ALBEDO

The amount of energy required to raise the temperature of one pound of water one degree Fahrenheit. — BRITISH THERMAL UNIT (BTU)

The amount of heat required to raise the temperature of one gram of water one degree Celsius. — CALORIE

The ratio of the energy collected by a solar collector to the radiant energy incident on the collector. — COLLECTOR EFFICIENCY

A material used to store heat by melting. Heat is later released for use as the material solidifies. — EUTECTIC SALT

Coal, oil, or natural gas. — FOSSIL FUEL

The heat absorbed by a body when undergoing a phase change from a solid to a liquid with no change in temperature. — HEAT OF FUSION

Heat loss by a body when undergoing a phase change from a liquid to a solid with no change in temperature (numerically equal to the heat of fusion). — HEAT OF SOLIDIFICATION

The angle between the direction of the sun and the perpendicular (normal) to the surface on which sunlight is falling. — INCIDENT ANGLE

Sunlight, or solar radiation, including wavelengths of ultraviolet (<0.4 microns), visible (0.4 microns to 0.7 microns), and infrared radiation (>0.7 microns). Total insolation includes both direct and diffuse insolation. See figure D–2, page 98, for spectral characteristics. — INSOLATION

A physical property of material indicating the amount of heat in calories or BTUs absorbed when a material changes from a solid to a liquid (fusion) or from a liquid to a gas (vaporization). Energy is given up when a gas condenses or when a liquid solidifies. — LATENT HEAT

The process involved when a material changes from a solid to a liquid or a liquid to a gas, each requiring an absorption of energy with no temperature change; or when the material changes from a gas to a liquid or a liquid to a solid, each requiring a loss of energy with no temperature change. — PHASE CHANGE

PHOTOVOLTAIC CELLS (Solar cells): Semiconducting devices that convert sunlight directly into electric power. The conversion process is called the photovoltaic effect.

PYRANOMETER An instrument for measuring sunlight intensity. It usually measures total (direct plus diffuse) insolation over a broad wavelength range.

PYRHELIOMETER An instrument that measures the intensity of the direct beam radiation (direct insolation) from the sun. The diffuse component is not measured.

SPECIFIC HEAT A physical property of materials that indicates the amount of heat required to raise the temperature of one pound of material one degree Fahrenheit, measured in BTU/lb °F; or one gram of material one degree Celsius measured in calorie/g °C.

Conversion Factors for
Solar Energy Study

1 m	= 100	cm
1 acre	= 43,560	ft²
1 acre	= 4,047	m²
1 BTU	= 252	cal
1 BTU	= 1,055	joules
1 BTU ft⁻²	= 0.271	langleys (cal cm⁻²)
1 cal	= 3.97×10⁻³	BTU
1 cal	= 3.09	ft-lb
1 cal	= 4.184	joules
1 cal min⁻¹	= 6.98×10⁻²	watts
1 cm	= 0.394	in
1 cc or cm³	= 6.10×10⁻²	in³
1 ft³	= 28.3	liters
1 in³	= 16.4	cc or cm³
1 ft	= 12	in
1 ft	= 0.305	m
1 ft-lb	= 0.324	cal
1 ft-lb	= 1.36	joules
1 gal	= 3.79	liters
1 g	= 2.20×10⁻³	lb
1 hp	= 0.745	kw
1 in	= 2.54	cm
1 joule	= 9.48×10⁻⁴	BTU
1 joule	= 0.239	cal
1 joule	= 0.738	ft-lb
1 kcal	= 3.97	BTU
1 kcal min⁻¹	= 6.98×10⁻²	kw
1 kg-m	= 7.23	ft-lb
1 kg	= 2.20	lb
1 kw	= 1.34	hp
1 kwh	= 3,413	BTU
1 kw	= 14.3	kcal min⁻¹
1 langley (cal cm⁻²)	= 3.69	BTU ft⁻²
1 langley min⁻¹ (cal/cm²·min)	= 6.98×10⁻²	watts cm⁻²
1 liter	= 0.264	gal
1 liter	= 1.06	qt
1 m	= 3.28	ft
1 m	= 39.4	in
1 m	= 6.21×10⁻⁴	miles
1 mile	= 1,609	m
1 lb	= 454	g
1 lb	= 0.454	kg
1 qt	= 0.946	liters

Appendix B

1 cm^2	=	1.08×10^{-3}	ft^2
1 cm^2	=	0.155	in^2
1 ft^2	=	929	cm^2
1 ft^2	=	144	in^2
1 ft^2	=	9.29×10^{-2}	m^2
1 in^2	=	6.45	cm^2
1 m^2	=	10.8	ft^2
1 m^2	=	3.86×10^{-7}	miles^2
1 mile^2	=	640	acres
1 mile^2	=	2.79×10^7	ft^2
1 mile^2	=	2.59×10^6	m^2
1 ton	=	907	kg
1 ton	=	2,000	lb
1 watt cm^{-2}	=	14.3	langleys (cal/cm$^2 \cdot$min)
1 yd	=	3	ft
1 yd	=	91.4	cm
1 cal cm^{-2} sec^{-1} C$^{\circ -1}$	=	7,380	BTU ft^{-2} hr^{-1} F$^{\circ -1}$
1 BTU ft^{-2} hr^{-1} F$^{\circ -1}$	=	1.35×10^{-4}	cal cm^{-2} sec^{-1} C$^{\circ -1}$

Temperature Scale Conversion

TEMPERATURE

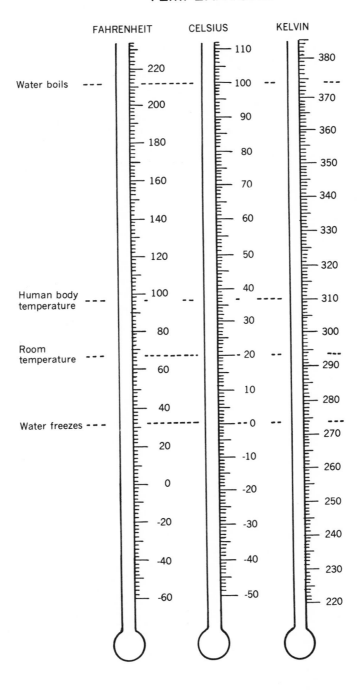

Conversion equations:

$$\frac{T_C}{100} = \frac{T_F - 32}{180}.$$

$$T_K = T_C + 273.$$

APPENDIX D

Solar Position and Intensity Data for 32°, 40°, and 48° North Latitude

Date	Solar Time A.M.*	32° N. Latitude			40° N. Latitude			48° N. Latitude		
		Solar Position		Total Insolation BTU/sq. ft. Rad. N.**	Solar Position		Total Insolation BTU/sq. ft. Rad. N.**	Solar Position		Total Insolation BTU/sq. ft. Rad. N.**
		Alt.	Azm.		Alt.	Azm.		Alt.	Azm.	
Jan. 21	7	1.4°	65.2°	1	. . .	. . .	. . .	. . .	. . .	. . .
	8	12.5	56.5	202	8.1°	55.3°	141	3.5°	54.6°	36
	9	22.5	46.0	269	16.8	44.0	238	11.0	42.6	185
	10	30.6	33.1	295	23.8	30.9	274	16.9	29.4	239
	11	36.1	17.5	306	28.4	16.0	289	20.7	15.1	260
	12	38.0	0.0	309	30.0	0.0	293	22.0	0.0	267
Feb. 21	7	6.7	72.8	111	4.3	72.1	55	1.8	71.7	3
	8	18.5	63.8	244	14.8	61.6	219	10.9	60.0	180
	9	29.3	52.8	287	24.3	49.7	271	19.0	47.3	247
	10	38.5	38.9	305	32.1	35.4	293	25.5	33.0	275
	11	44.9	21.0	314	37.3	18.6	303	29.7	17.0	288
	12	47.2	0.0	316	39.2	0.0	306	31.2	0.0	291
Mar. 21	7	12.7	81.9	184	11.4	80.2	171	10.0	78.7	152
	8	25.1	73.0	260	22.5	69.6	250	19.5	66.8	235
	9	36.8	62.1	289	32.8	57.3	281	28.2	53.4	270
	10	47.3	47.5	304	41.6	41.9	297	35.4	37.8	287
	11	55.0	26.8	310	47.7	22.6	304	40.3	19.8	295
	12	58.0	0.0	312	50.0	0.0	306	42.0	0.0	297
Apr. 21	6	6.1	99.9	66	7.4	98.9	89	8.6	97.8	108
	7	18.8	92.2	206	18.9	89.5	207	18.6	86.7	205
	8	31.5	84.0	256	30.3	79.3	253	28.5	74.9	247
	9	43.9	74.2	278	41.3	67.2	275	37.8	61.2	269
	10	55.7	60.3	290	51.2	51.4	286	45.8	44.6	281
	11	65.4	37.5	296	58.7	29.2	292	51.5	24.0	287
	12	69.6	0.0	298	61.6	0.0	294	53.6	0.0	289
May 21	5	. . .	. . .	. . .	1.9°	114.7°	1	5.2°	114.3°	41
	6	10.4°	107.2°	118	12.7	105.6	143	14.7	103.7	162
	7	22.8	100.1	211	24.0	96.6	216	24.6	93.0	218
	8	35.4	92.9	249	35.4	87.2	249	34.6	81.6	248
	9	48.1	84.7	269	46.8	76.0	267	44.3	68.3	264
	10	60.6	73.3	279	57.5	60.9	277	53.0	51.3	274
	11	72.0	51.9	285	66.2	37.1	282	59.5	28.6	279
	12	78.0	0.0	286	70.0	0.0	284	62.0	0.0	280

Source: Reprinted by permission from *ASHRAE Handbook of Fundamentals*, 1972.

Note: Make appropriate interpolations for the latitude of your own region.

* P.M. values mirror the A.M. values; 8 A.M.–4 P.M., 9 A.M.–3 P.M., 10 A.M.–2 P.M., etc.

** Radiation on a normal surface, representative of conditions on average cloudless days. The water content of the atmosphere can cause this to vary as much as 40%.

Date	Solar Time A.M.*	32° N. Latitude			40° N. Latitude			48° N. Latitude		
		Solar Position		Total Insolation BTU/sq. ft.	Solar Position		Total Insolation BTU/sq. ft.	Solar Position		Total Insolation BTU/sq. ft.
		Alt.	Azm.	Rad. N.**	Alt.	Azm.	Rad. N.**	Alt.	Azm.	Rad. N.**
Jun. 21	5	. . .	. . .	. .	4.2	117.3	21	7.9	116.5	77
	6	12.2°	110.2°	130	14.8	108.4	154	17.2	106.2	172
	7	24.3	103.4	209	26.0	99.7	215	27.0	95.8	219
	8	36.9	96.8	244	37.4	90.7	246	37.1	84.6	245
	9	49.6	89.4	263	48.8	80.2	262	46.9	71.6	260
	10	62.2	79.7	273	59.8	65.8	272	55.8	54.8	269
	11	74.2	60.9	278	69.2	41.9	276	62.7	31.2	273
	12	81.5	0.0	280	73.5	0.0	278	65.4	0.0	275
Jul. 21	5	. . .	. . .	. . .	2.3	115.2	2	5.7	114.7	42
	6	10.7	107.7	113	13.1	106.1	137	15.2	104.1	155
	7	23.1	100.6	203	24.3	97.2	208	25.1	93.5	211
	8	35.7	93.6	241	35.8	87.8	241	35.1	82.1	240
	9	48.4	85.5	261	47.2	76.7	259	44.8	68.8	256
	10	60.9	74.3	271	57.9	61.7	269	53.5	51.9	266
	11	72.4	53.3	277	66.7	37.9	274	60.1	29.0	271
	12	78.6	0.0	278	70.6	0.0	276	62.6	0.0	272
Aug. 21	6	6.5	100.5	59	7.9	99.5	80	9.1	98.3	98
	7	19.1	92.8	189	19.3	90.0	191	19.1	87.2	189
	8	31.8	84.7	239	30.7	79.9	236	29.0	75.4	231
	9	44.3	75.0	263	41.8	67.9	259	38.4	61.8	253
	10	56.1	61.3	275	51.7	52.1	271	46.4	45.1	265
	11	66.0	38.4	281	59.3	29.7	277	52.2	24.3	271
	12	70.3	0.0	283	62.3	0.0	279	54.3	0.0	273
Sep. 21	7	12.7°	81.9°	163	11.4°	80.2°	149	10.0°	78.7°	131
	8	25.1	73.0	240	22.5	69.6	230	19.5	66.8	215
	9	36.8	62.1	272	32.8	57.3	263	28.2	53.4	251
	10	47.3	47.5	287	41.6	41.9	279	35.4	37.8	269
	11	55.0	26.8	294	47.7	22.6	287	40.3	19.8	277
	12	58.0	0.0	296	50.0	0.0	290	42.0	0.0	280
Oct. 21	7	6.8	73.1	98	4.5	72.3	48	2.0	71.9	3
	8	18.7	64.0	229	15.0	61.9	203	11.2	60.2	165
	9	29.5	53.0	273	24.5	49.8	257	19.3	47.4	232
	10	38.7	39.1	292	32.4	35.6	280	25.7	33.1	261
	11	45.1	21.1	301	37.6	18.7	290	30.0	17.1	274
	12	47.5	0.0	304	39.5	0.0	293	31.5	0.0	278
Nov. 21	7	1.5	65.4	. . .	. . .	. . .	. . .	. . .	. . .	. . .
	8	12.7	56.6	196	8.2	55.4	136	3.6	54.7	36
	9	22.6	46.1	262	17.0	44.1	232	11.2	42.7	178
	10	30.8	33.2	288	24.0	31.0	267	17.1	29.5	232
	11	36.2	17.6	300	28.6	16.1	283	20.9	15.1	254
	12	38.2	0.0	303	30.2	0.0	287	22.2	0.0	260
Dec. 21	8	10.3	53.8	176	5.5	53.0	88	. . . .	. . .	. .
	9	19.8	43.6	257	14.0	41.9	217	8.0	40.9	140
	10	27.6	31.2	287	20.7	29.4	261	13.6	28.2	214
	11	32.7	16.4	300	25.0	15.2	279	17.3	14.4	242
	12	34.6	0.0	304	26.6	0.0	284	18.6	0.0	250

APPENDIX E
Thermal Resistances of Building Materials

		R for thickness listed
Air Spaces	Air bounded by ordinary materials, vertical and ¾″ or wider	0.97
Surfaces	Still air, vertical, inside	0.68
	15-mph wind speed, outside	0.17
	Still air, horizontal, inside	0.61
	Still air, 45° slope, inside	0.62
Exterior Finishes	Brick veneer, 4″ thick (nominal)	0.44
	Wood shingles, 16″ with 7½″ exposure	0.93
	Asbestos-cement shingles	0.27
	Wood siding, bevel ½″ × 8″, lapped	0.81
	Aluminum, hollow-backed over sheathing	0.61*
	Building paper, permeable felt	0.06

*This is the resistance of the air film only. Aluminum has negligible resistance.

Masonry Materials	Brick, common, 4″ thick	0.80
	Concrete block, 3-oval core, sand and gravel aggregate, 8″	1.11
	Concrete, sand, and gravel or stone aggregate, per inch of thickness	0.08
	Stone, per inch of thickness	0.08
Wood	Hardwoods, per inch of thickness	0.91
	Softwoods and plywood, per inch of thickness	1.25
Windows	(Treat doors, with or without glass, the same as windows.)	
	Single (no storm sash)	0.89
	With storm sash	1.89
	Double glazed with ¼″ air space	1.64
Building Board and Paper	Gypsum or plasterboard, ⅜″ thick	0.32
	Gypsum or plasterboard, ½″ thick	0.45
	Vapor-permeable felt paper	0.06

	R for thickness listed	
Insulating board (wood or cane fiber), $^{25}\!/_{32}''$	2.06	Sheathing
Plywood, $^{1}\!/_{4}''$	0.39	
Fir or pine, $^{25}\!/_{32}''$	0.98	
Insulating board, $^{1}\!/_{2}''$ asphalt impregnated	1.32	
Gypsum board, $^{3}\!/_{8}''$	0.32	Interior
Gypsum board, $^{1}\!/_{2}''$	0.45	Finishes
Insulating board, $^{1}\!/_{2}''$	1.43	
Metal lath and plaster (gypsum, sand aggregate), $^{3}\!/_{4}''$ plaster	0.13	
Blanket or bat		Insulating
Wood fiber, per inch	4.00	Materials
Mineral wool or fiberglass, per inch	3.70	
Loose fill		
Mineral wool or fiberglass, per inch	3.33	
Vermiculite (expanded), per inch	2.08	
Perlite (expanded), per inch	2.70	
Board		
Fiberglass, per inch	4.00	
Asphalt shingles	0.44	Roofing
Wood shingles	0.94	
Slate, $^{1}\!/_{2}''$	0.05	

To determine the rate of heat transfer through a wall or roof, add the resistances of individual components. Determine the reciprocal of total resistance. This is the U value for rate of heat transfer in $BTU/(hr)(ft^2)(°F)$.

Example:

Construction of wall	Resistance R
1. Outside surface (15-mph wind)	0.17
2. Wood siding, $^{1}\!/_{2}'' \times 8''$ lapped	0.81
3. Building paper, permeable felt	0.06
4. Sheathing, insulating board, $^{1}\!/_{2}''$ asphalt impregnated	1.32
5. Air space	0.97
6. Gypsum wallboard	0.45
7. Inside surface (still air)	0.68
Total R	4.46

$$U = \frac{1}{R} = \frac{1}{4.46} = 0.234 \ BTU/(hr)(ft^2)(°F).$$

To adjust the U value for wall area occupied by studs, compute U for wall at studs. Find 15% of U for stud area and 85% of U for wall area. Add these values for net U for the wall.

Construction of 45° pitched roof (heat flow up)	Resistance R
1. Outside surface (15-mph wind)	0.17
2. Wood shingles	0.94
3. Roof boards, nominal 1″ fir × 6″ with 3″ space between	0.70(approx.)
4. Fiberglass insulation, 6″ bats	22.20
5. Gypsum wallboard, ½″	0.45
6. Inside surface (still air)	0.62
Total R	25.08

$$U = \frac{1}{R} = \frac{1}{25.08} = 0.04 \text{ BTU}/(\text{hr})(\text{ft}^2)(°\text{F}).$$

Adjust U for roof area occupied by rafters in a manner similar to that for wall studs.

Note: For more complete treatment of heat transfer situation involving buildings, refer to:

1. *ASHRAE Handbook of Fundamentals,* 1972.
2. *Heat Loss Calculation Guide H-21,* Institute of Boiler and Radiator Manufacturers.

Reprinted by permission from *ASHRAE Handbook of Fundamentals,* 1972.

Table of Density of Air Based on Various Conditions of Temperature and Pressure

Temperature in Degrees Celsius	Pressure in Inches of Mercury	Density in Grams per Liter	Temperature in Degrees Celsius	Pressure in Inches of Mercury	Density in Grams per Liter
0	28.80	1.239	18	28.80	1.163
0	29.20	1.256	18	29.20	1.179
0	29.60	1.274	18	29.60	1.195
0	30.00	1.291	18	30.00	1.211
0	30.40	1.308	18	30.40	1.227
0	30.80	1.325	18	30.80	1.243
3	28.80	1.226	21	28.80	1.151
3	29.20	1.243	21	29.20	1.167
3	29.60	1.260	21	29.60	1.183
3	30.00	1.277	21	30.00	1.199
3	30.40	1.294	21	30.40	1.215
3	30.80	1.311	21	30.80	1.231
6	28.80	1.212	24	28.80	1.139
6	29.20	1.229	24	29.20	1.155
6	29.60	1.246	24	29.60	1.171
6	30.00	1.263	24	30.00	1.187
6	30.40	1.280	24	30.40	1.202
6	30.80	1.297	24	30.80	1.218
9	28.80	1.200	27	28.80	1.128
9	29.20	1.216	27	29.20	1.143
9	29.60	1.233	27	29.60	1.159
9	30.00	1.250	27	30.00	1.175
9	30.40	1.266	27	30.40	1.190
9	30.80	1.283	27	30.80	1.206
12	28.80	1.187	30	28.80	1.117
12	29.20	1.203	30	29.20	1.132
12	29.60	1.220	30	29.60	1.148
12	30.00	1.236	30	30.00	1.163
12	30.40	1.253	30	30.40	1.179
12	30.80	1.269	30	30.80	1.194
15	28.80	1.175	33	28.80	1.106
15	29.20	1.191	33	29.20	1.121
15	29.60	1.207	33	29.60	1.136
15	30.00	1.224	33	30.00	1.152
15	30.40	1.240	33	30.40	1.167
15	30.80	1.256	33	30.80	1.182

Temperature in Degrees Celsius	Pressure in Inches of Mercury	Density in Grams per Liter	Temperature in Degrees Celsius	Pressure in Inches of Mercury	Density in Grams per Liter
36	28.80	1.095	48	30.00	1.098
36	29.20	1.110	48	30.40	1.112
36	29.60	1.125	48	30.80	1.127
36	30.00	1.140			
36	30.40	1.156	51	28.80	1.044
36	30.80	1.171	51	29.20	1.059
			51	29.60	1.073
39	28.80	1.084	51	30.00	1.088
39	29.20	1.099	51	30.40	1.102
39	29.60	1.114	51	30.80	1.117
39	30.00	1.129			
39	30.40	1.145	54	28.80	1.035
39	30.80	1.160	54	29.20	1.049
			54	29.60	1.063
42	28.80	1.074	54	30.00	1.078
42	29.20	1.089	54	30.40	1.092
42	29.60	1.104	54	30.80	1.106
42	30.00	1.119			
42	30.40	1.134	57	28.80	1.025
42	30.80	1.149	57	29.20	1.039
			57	29.60	1.054
45	28.80	1.064	57	30.00	1.068
45	29.20	1.079	57	30.40	1.082
45	29.60	1.093	57	30.80	1.096
45	30.00	1.108			
45	30.40	1.123	60	28.80	1.016
45	30.80	1.138	60	29.20	1.030
			60	29.60	1.044
48	28.80	1.054	60	30.00	1.058
48	29.20	1.069	60	30.40	1.072
48	29.60	1.083	60	30.80	1.087

Suggestions for Further Reading

Selected Bibliography on Solar Energy

Abetti, Georgio. *THE SUN.* New York: Macmillan Co., 1957.

Alfven, Hannes. *ON THE ORIGIN OF THE SOLAR SYSTEM.* Westport, Connecticut: Greenwood Press, 1973.

Berlage, H. P. *THE ORIGIN OF THE SOLAR SYSTEM.* Flushing, New York: Pergamon Press, 1968.

Blanco, V. M., and McCuskey, S. W. *BASIC PHYSICS OF THE SOLAR SYSTEM.* Reading, Massachusetts: Addison-Wesley, 1961.

Bonestell, Chesley. *SOLAR SYSTEM.* Chicago: Childrens Press, 1968.

Bonner, Mary G. *WONDERS AROUND THE SUN.* New York: Lantern Books.

Branley, Franklyn M. *SOLAR ENERGY.* New York: Thomas Y. Crowell, 1957.

Brinkworth, B. J. *SOLAR ENERGY FOR MAN.* New York: Halsted Press, 1973.

Cook, James G. *WE LIVE BY THE SUN.* New York: Dial Press, 1957.

Daniels, Farrington. *DIRECT USE OF THE SUN'S ENERGY.* New Haven: Yale University Press, 1964.

Daniels, Farrington, and Duffie, John. *SOLAR ENERGY RESEARCH.* Madison: University of Wisconsin Press, 1955.

Duffie, John, and Beckman, William. *SOLAR ENERGY THERMAL PROCESSES.* New York: John Wiley & Sons, 1974.

Gamow, George. *THE BIRTH AND DEATH OF THE SUN.* New York: New American Library, 1955.

Halacy, Dan S. *THE COMING AGE OF SOLAR ENERGY.* New York: Harper & Row, 1973.

————. *EXPERIMENTS WITH SOLAR ENERGY.* New York: Grossett & Dunlap.

————. *FABULOUS FIREBALL.* New York: Macmillan Co., 1957.

————. *FUN WITH THE SUN.* New York: Scholastic Book Service, 1972.

Hoke, John. *FIRST BOOK OF SOLAR ENERGY.* New York: Franklin Watts, 1968.

International Rectifier Company. *SOLAR CELLS AND PHOTOCELLS,* 1973. Semiconductor Di, 233 Kansas Street, El Segundo, California 90245.

Kiepenheuer, Karl. *THE SUN.* Ann Arbor: University of Michigan Press, 1959.

Rau, Hans. *SOLAR ENERGY.* New York: Macmillan Co., 1964.

Robinson, N. *SOLAR RADIATION.* New York: American Elsevier, 1966.

Ruchlis, Hy. *THANK YOU, MR. SUN.* New York: Harvey House, 1957.

Simak, Clifford D. *SOLAR SYSTEM: OUR NEW FRONT YARD.* New York: St. Martin's Press, 1962.

Ubbelohde, A. R. *MAN AND ENERGY.* Baltimore, Maryland: Penguin Books, Pelican, 1963.

Wohlrabe, Raymond. *EXPLORING SOLAR ENERGY.* Cleveland, Ohio: World, 1966.

Zarem, A. M., and Erway, D. D. *INTRODUCTION TO THE UTILIZATION OF SOLAR ENERGY.* New York: McGraw-Hill, 1963.

ACKNOWLEDGMENTS

The authors wish to express their thanks to the many students whose experimental efforts contributed significantly to this manual. Included in this list are Keith Dayer, Michael Jewett, Eliot Salant, Emily Sherwood, Rolf Rogers for his line drawings, and in particular Alex Pidwerbetsky. Appreciation is also extended to the Mid-Atlantic States Air Pollution Control Association for their kind permission to use the Atmospheric Turbidity, Volz Sun Photometer, and Solar Energy Use experiments from their *Air Pollution Experiments for Junior and Senior High School Science Classes*. Special thanks go to the staff of the Atmospheric Sciences Research Center for their many helpful suggestions and improvements to the manuscript. Our gratitude also goes to Linda Sikorowski, Assistant to the Director, ASRC, for her administrative and editorial assistance.

Index

The letter t in parentheses ( ) following a page number indicates a table. An italicized page number indicates a figure.